GENETICS 101

GENETICS 101

MICHAEL WINDELSPECHT

SCIENCE 101

GREENWOOD PRESS
Westport, Connecticut • London

Library of Congress Cataloging-in-Publication Data

Windelspecht, Michael, 1963–
Genetics 101/Michael Windelspecht.
  p. cm. – (Science 101, ISSN 1931–3950)
Includes bibliographical references.
ISBN 978–0–313–33381–1 (alk. paper)
1. Genetics–Popular works.   I. Title.
QH437.W56 2007
576.5–dc22   2007014304

British Library Cataloguing in Publication Data is available.

Library of Congress Catalog Card Number: 2007014304
ISBN-13: 978–0–313–33381–1
ISBN-10: 0–313–3381–5
ISSN: 1931-3950

First published in 2007

Greenwood Press, 88 Post Road West, Westport, CT 06881
An imprint of Greenwood Publishing Group, Inc.
www.greenwood.com

Printed in the United States of America

The paper used in this book complies with the
Permanent Paper Standard issued by the National
Information Standards Organization (Z39.48–1984).

10 9 8 7 6 5 4 3 2 1

To my children, Devin and Kayla,
may your love of books endure

Contents

SERIES FOREWORD

What should you know about science? Because science is so central to life in the twenty-first century, science educators believe that it is essential that *everyone* understand the basic foundations of the most vital and far-reaching scientific disciplines. *Science 101* helps you reach that goal—this series provides readers of all abilities with an accessible summary of the ideas, people, and impacts of major fields of scientific research. The volumes in the series provide readers—whether students new to the science or just interested members of the lay public—with the essentials of a science using a minimum of jargon and mathematics. In each volume, more complicated ideas build upon simpler ones, and concepts are discussed in short, concise segments that make them more easily understood. In addition, each volume provides an easy-to-use glossary and an annotated references and resources of the most useful and accessible print and electronic resources that are currently available.

Acknowledgments

I would like to start off by thanking Greenwood Press for the opportunity to produce this volume of the Science 101 series. Since the time I received my doctorate in biology I have wanted to write a book for nonscientists which introduced them to the science of genetics. These types of books are important if we are to ever bridge the gap between the scientific community and the general public. I would also like to thank my editor, Kevin Downing. His unbelievable patience and endurance is greatly appreciated.

My wife Sandy has played an important role in my career as an author. She encourages my undertakings while doubling as my illustrator. In that regard she has also proven to be patient, as many of my initial concepts of the illustrations in this book have had to undergo multiple revisions before being finalized. Everything I am as a writer and scientist I owe to her.

ABBREVIATIONS

A	adenine
ADA	adenosine deaminase deficiency
AIDS	acquired immunodeficiency syndrome
BCE	before the common era
bp	base pair
Bt	*Bacillus thuringiensis*
C	cytosine
cDNA	complementary DNA
cM	centiMorgan
ddNTP	dideoxynucleotide triphosphate
DDT	dichloro-diphenyl-trichloroethane
DNA	deoxyribonucleic acid
dsDNA	double-stranded DNA
dsRNA	double-stranded RNA
EMS	ethyl methanesulfonate
ES	embryonic stem cells
FASPS	familial advanced sleep phase syndrome
FISH	fluorescent in situ hybridization
G	guanine
GMO	genetically modified organism
HIV	human immunodeficiency
HLA	human leukocyte antigen
IDDM	insulin-dependent diabetes mellitus
kb	kilobases
mb	megabases
MDD	major depression disorder
MHC	major histocompatability complex
mRNA	messenger RNA

mtDNA	mitochondrial DNA
mu	map unit
NIDDM	non-insulin-dependent diabetes mellitus
OTC	ornithine transcarbamylase deficiency
PCR	polymerase chain reaction
RFLP	restriction fragment length polymorphism
RISC	RNA-induced silencing complexes
RNA	ribonucleic acid
RP	retinitis pigmentosa
rRNA	ribosomal RNA
RNAi	RNA interference
SCID	severe combined immunodeficiency disease
siRNA	small interfering RNA
SNP	single-nucleotide polymorphism
snRNA	small nuclear RNA
ssDNA	single-stranded DNA
SSR	simple sequence repeat
ssRNA	single-stranded RNA
STR	short tandem repeat
STS	sequence-tagged site
T	thymine
tRNA	transfer RNA
U	uracil
UV	ultraviolet
VNTR	variable number of tandem repeats
XP	xeroderma pigmentosa

INTRODUCTION

Genetics is the study of inheritance. It is a broad science that examines the molecular basis of inheritance at the cellular level, the transmission of traits from generation to generation, and the movement of genes within and between populations. As a science, genetics is both very young and very old. It has only been in the past 50 years that we have discovered the structure of the molecule that carries our genetic information from generation to generation. That molecule is of course deoxyribonucleic acid, or DNA. Yet despite the fact that we have only been studying DNA for a short period of time, the study of genetics now permeates all aspects of the scientific world and our lives. This is evident by the fact that on any given day, newspapers and news Web sites contain articles discussing the latest advances in the field of genetics. A little more then a decade ago, scientists were debating the feasibility of cloning a mammal, now the discussion focuses on whether researchers should be allowed to clone humans. In a fifty-year time span we have learned the role of genes as the carriers of genetic information, and developed the ability to manipulate those genes to produce genetically modified and transgenic organisms.

Genetics is also a very old science. Contrary to what is frequently taught in science classrooms, the science of genetics did not begin with the work of Gregor Mendel in the nineteenth century. The principles of genetics have been in practice since humans first domesticated the dog, cow, chicken, and goat. Without an understanding of the methods, farmers of early civilizations recognized that specific traits could be selected for and in the process they could develop true-breeding lines of certain animals. The same principles applied to plants. Maize, corn, and rice are all examples of early genetic engineering by ancient civilizations. By cross-breeding and selecting for plants with desirable traits, farmers were able to establish food crops that produced larger fruits (and seeds) and made harvesting easier. Civilizations that could produce more food could also sustain larger armies. The cultural influence of these civilizations expanded over time, and with their success spread their knowledge of selecting for favorable traits in agriculture. Thus, indirectly these early geneticists established the foundations for modern society. Early genetics was not limited solely to the study of food crops and domesticated animals. In many early cultures, including those that were the precursors of some of today's countries, the priests and royal families kept detailed records of lineages and bloodlines in order to establish a hierarchy of power. In the process, they also discovered that some traits followed patterns. One of these, sex-linked inheritance, is hinted at in the historical records of several ancient civilizations.

Despite the historical contributions of genetic scientists to the development of scientific thought and modern importance of genetic research in the fields of both agriculture and medicine, most people still view the work of geneticists as some form of magic that is beyond their capability to understand. However, when provided with nontechnical discussions of genetics, people usually realize that many of the discoveries in genetics are logical, relatively easy to understand, and even interesting. The purpose of this book is to develop a level of appreciation for the science of genetics, with the hope that the reader will then search out additional sources to explore their particular interests in the study of inheritance.

One of the concepts that must be recognized is that the scientists who are involved in studying genetics are often very different from one another. Geneticists may work at a variety of levels. Molecular geneticists, cytogeneticists, cell biologists, and biochemists all study the DNA molecule and the proteins that are encoded by its information strands. These individuals typically work at the level of the cell. Transmission geneticists also study DNA, but focus on the examination of patterns of inheritance between individuals. Genetic counselors are professionals that also participate at this level of study. At the level of the individual genetics begins to include a more mathematical approach as the scientists involved utilize the principles of probability and frequency to describe the chances of inheritance. The next level of complexity involves the study of the movement of traits and alleles between populations. This is the work of the population geneticist. Population geneticists are interested in gene flow, immigration and emigration, the effects of disruptions in populations and how these events influence the frequency of a trait in a given group of individuals. At the broadest level, evolutionary geneticists and genomicists examine how the sum of the genes in a species, also called the *genome*, changes over long periods of time. The past decade has seen a tremendous focus on this level of genetics as new technologies and computer applications have allowed the generation and study of large databases of genetic information.

Regardless of their level of focus, all geneticists are involved with the study of the gene. Simply put, genes contain the instructions necessary for the expression of a trait in an individual. Typically the term *trait* is used to explain an observable characteristic of an organism. The physical trait of coat color in rabbits is an example. However, many traits are involved with the complicated biochemical pathways of an organism. Alcohol production by yeast, antibiotic resistance in bacteria, and lactose intolerance in humans are all examples of physiological and biochemical traits that are not easily detected by the appearance of the organism. One of the objectives of this book will be to help the reader understand some of the processes by which scientists study the gene and learn of its function in an organism by manipulating genes in model organisms.

Genetics is also not an isolated science. The study of genetics has been made possible by the countless contributions of biochemists, cell biologists, chemists, and physicists. In addition, psychologists who study behavior have yielded important insights on the interactions of the environment and the gene. The field of behavioral genetics, which has direct application in the pharmaceutical industries, is a result of these interdisciplinary forms of research. Perhaps one of the biggest interdisciplinary contributors is the field of computer science. Computer scientists have made enormous contributions toward our understanding of the gene by providing software for the large genetic databases, modeling studies, and the rapid analysis of genetic sequences. A visitor at any molecular genetics lab will often find that researchers spend as much time sitting at a computer as they do working at a research bench. For example, once a molecular geneticist has generated a sequence of DNA from an automated sequencer (which contains elaborate computer software itself), they often utilize complex algorithms and statistical packages to examine the structure of the gene they are studying. In addition, the development of the supercomputer and the Internet has made it possible for geneticists to analyze their data quickly and over vast differences. Almost all genetics labs possess computers that are networked into large central databases of genetic information. Without these advances, projects such as the Human Genome Project would not have been possible.

While genetics has received input from a variety of disciplines, it also contributes immensely to our understanding of the natural world around us. Many of these contributions are addressed in the entries of this work. One of the biggest areas of contribution is in the field of medicine. Not only has genetics allowed for us to predict the probability of getting a specific disease based upon our inherited set of genes, it has enabled the synthesis of drugs designed specifically to regulate gene expression. These drugs are frequently produced by transgenic organisms, another byproduct of the genetic revolution.

In the past decade some of the greatest advances in genetics have been made by the plant geneticists. By using model organisms, such as *Arabidopsis*, plant geneticists have developed an understanding of plant water and nutrient use that have allowed them to develop strains of food-producing species that can survive in drought conditions and nutrient-poor soils. This has greatly benefited our ability to produce food in areas of the world plagued by starvation. Plant geneticists have also developed techniques to produce transgenic and genetically modified plant species that are resistant to disease and insect pests. This has served to increase crop yields around the globe, another benefit for a planet with an expanding human population.

The science of genetics is also beginning to provide some answers to long-debated questions on human evolution. Through an improvement of techniques, scientists are now able to extract DNA from our extinct relatives, including the Neanderthals. By comparing this DNA with our own and our close primate relatives, geneticists are beginning to zero in on those genes which enable cognitive thoughts and physiological changes known to be important distinctions between us and our primate cousins. In the near future we may be able to more fully understand what makes humans unique among the animals.

One of the first purposes of this work is to examine how scientists determined that traits are coded for by some physical structure, which we now call the gene. The first chapter begins with a brief overview of the history of genetics prior to the nineteenth century. This is followed by a discussion of how early geneticists recognized that traits obeyed (for the most part) mathematical laws. This begins the study of classical (or transmission) genetics and involves the work of individuals such as Gregor Mendel and Thomas Hunt Morgan. Also in this chapter is an entry which defines key genetic terms that are important for an understanding of the remainder of the book. Chapter 2 focuses on how scientists determined that deoxyribonucleic acid, or DNA, contains the genes. In today's world this may seem to be common knowledge, but in the early part of the twentieth century many scientists actually believed that proteins were the hereditary material. Only through a series of experiments did the scientific community develop enough evidence to establish DNA's central role as the carrier of the information on which genes are found.

Having established that DNA contains genes, Chapter 3 examines the structure of a gene, and how this information it stored and utilized by the cell. The processes of transcription and translation are covered in this chapter, as well as an introduction to DNA replication. The study of DNA replication has special importance for geneticists. Once scientists determined the process by which a cell copies its DNA it became possible to mimic this procedure in a test tube. The end result was the invention of the polymerase chain reaction, or PCR—a procedure that has revolutionized the study of genetics.

Chapter 4 introduces some of the ways by which scientists study the gene. This chapter represents an overview of molecular genetics, and provides entries on PCR, how scientists clone genes, and the technology associated with the production of recombinant DNA. It also discusses many of the procedures that are seen on television programs involving forensic scientist. Chapter 5 examines how the genetic material may change as the result of mutation or chromosomal abnormalities. This chapter also introduces transposons, or "jumping genes". These are mobile genetic elements that appear to play an important role in evolutionary genetics.

The final chapters of the book examine some of the more recent developments in the study of the gene. In Chapter 6 we examine how scientists manipulate a gene or the genome of an organism. Genes do not operate in a vacuum. In order to understand gene function it is often first necessary to introduce mutations into the sequence of nucleotides and then reintroduce the gene into a model organism to determine the effect at the level of the organism. Scientists had a variety of tools with which to perform this task. This chapter also examines some relatively recent advances, such as RNA interference (RNAi) and gene therapy. In the final chapter of the book we take a look at some of the applications of genetics and some of the frontiers of genetic research. This chapter includes a discussion of DNA typing, also known as DNA fingerprinting, a common theme for many science-related forensic shows on television. In addition, this chapter will take a look at stem cells and the cloning of an organism, including the possibilities of cloning a human.

It is important to note that this work is not a textbook of genetics. Genetics represents an immense field of study in the life sciences, and it is simply not possible to cover all aspects of genetics in the scope of this work. Furthermore, most genetics textbooks assume that the reader has a background in the sciences, specifically in the fields of biology and chemistry. Such is not the case with this book. Instead, this work has been designed as an initial reference source for individuals who have an interest in genetics, but are not scientists. Throughout the book scientific jargon is kept to a minimum. However, as is the case with most scientific disciplines, genetics has its own language. To facilitate an understanding of this language, key terms are defined within the book, and a Glossary is provided at the end of the book. Terms that listed in the glossary are **boldfaced** at their first introduction in the book. A complete list of references is provided within the References and Resources chapter. These references are suitable for individuals who wish to obtain detailed information on some aspect of the material covered in this work. Many of these references are designed for nonscientists. Since the study of genetics has been revolutionized by the rapid access to information on the Internet, a list of key Web sites is provided in this chapter as well. Many of these Web sites act as portals into an ever-expanding resource of genetic information on the web. An index is also provided so that the reader may cross-refer concepts as needed. The illustrations and tables within this volume have been chosen to enhance the reader's understanding of the concepts. In many cases they have been simplified to provide an overview of a genetic principle.

The nature of this book makes it a useful reference for secondary school libraries, undergraduate higher education colleges, and universities where students may be seeking general information on some aspect of genetics. In addition, community libraries that wish to possess a general reference volume on genetics, as well as anyone with an interest in this field of study, will find this a useful addition to their collection.

1

GENETICS AS A SCIENCE

CORN AND WHEAT: BIRTH OF CIVILIZATION AND GENETICS

It is mistakenly believed by many that the study of genetics did not start until the nineteenth century. While it is true that discoveries in the nineteenth century played a major role in the development of modern genetics, especially with regards to the application of mathematics and statistics to the study of inheritance, the roots of genetics as a science reach far into our past, back to the very dawn of civilization. Archeologists have uncovered evidence of genetic experiments that date back over 12,000 years. In fact, we are all very familiar with the experiments of these early "geneticists." They not only form the foundation of our food supply, they also serve humans as pets and animals of agriculture. Most historians agree that without these advances modern civilization would not have developed into what we experience today.

Bread wheat (*Triticum aestivum*) is one of the staples of modern agriculture. It is responsible for feeding a large portion of the world's **population** daily and by some estimates provides over 20 percent of the daily calorie intake for the human species. However, the plant we recognize as wheat today is not really "natural," it is the result of thousands of years of artificial selection by early farmers.

Wheat belongs to the grass family of plants. If you take a walk through a mature field of grass late in the summer you can observe thousands of wheat-relatives that are putting out seed for the next generation. Grass plants, like most living organisms, do not mature at the same time, which allows the plant to disperse seeds over an extended period, thus increasing the chances that a percentage of the seeds will find favorable weather and soil conditions for surviving. Grass seeds also tend to be small and have a limited nutritional value. Thus, early wheat-like grasses would not produce a sufficient harvestable yield to support even a small group of people, much less an entire society. However, if by selecting for plants with larger seeds, for plants that matured at approximately the same time, and for plants whose seeds could be harvested at the same time, it would become possible to effectively use this plant as a mainstay of your diet. Furthermore, the use of the wheat would anchor your small group to a specific region, effectively beginning the formation of farming communities.

Obviously this is what happened, but it took some time for the selection process to produce modern bread wheat. By studying the **chromosomes** of modern wheat, and comparing them to other members of the wheat family, agricultural geneticists have learned that early farmers effectively selected for specific traits that benefited harvest. Figure 1.1 demonstrates the basic crosses that produced modern bread wheat. The final cross in the series, between Emmer wheat and the wild grass *Triticum tauschi* (producing bread wheat), is believed to have first occurred around 5000–6000 BCE in the area now occupied by countries such as Iraq, Iran, and Syria. Of course, these early people did not know about deoxyribonucleic acid (DNA) or genetics, but they did know how to carefully select for specific traits. Charles Darwin (1809–1882) called this process artificial selection, and used it as the basis of his development of the theory of natural selection in the nineteenth century. Since traits, such as seed size, are determined by **genes**, the production of wheat represents one of the earliest experiments in genetics.

While wheat was being domesticated in Asia, the native populations of Central America were busy selecting for favorable traits in maize, or corn (*Zea mays*) as it is commonly called. Like wheat, corn is a derivative of a grass. In this case the grass ancestor is teosinte (*Zea mays parviglumis*), a grass that is native to the area now known as southern Mexico. The domestication of corn in the New World played the same role as wheat in the Old World, it allowed for a cultural transition to a stationary agricultural society.

Recent studies of DNA from corn, and its close relatives, indicates that the process of domestication began around 4250 BCE, although some researchers believe that this date may be eventually pushed back to around 7000 BCE as additional studies are conducted. The domestication of corn by selecting for favorable traits was mostly complete before the plant was introduced into North America (about 3,200 years ago), and subsequently the remainder of the world. This suggests a relatively narrow time frame, maybe less than 1,000 years, during which early corn farmers began to select for three traits that distinguish

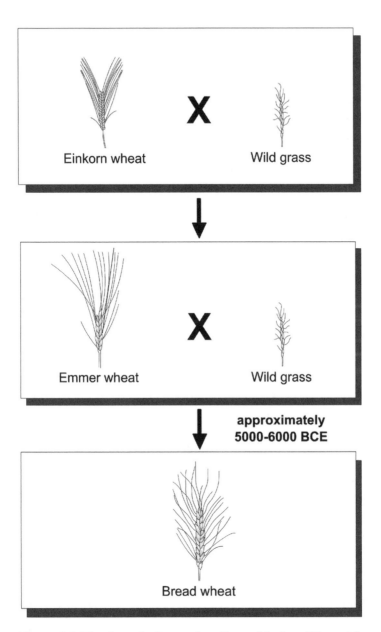

Figure 1.1 The Genetic Crosses Leading to Modern Wheat (*Courtesy of Ricochet Productions*).

modern maize from teosinte. These traits determine the basic structure of the plant, proteins associated with seed storage, and **enzymes** involved in the starch structure of the plant. Selection for the first of these traits, plant structure, is what gives corn its characteristic tall, narrow form. In addition, the selection for plant structure involved choosing plants in which the female structures, commonly called the cobs, are located near the center line of the plant. These changes in structure greatly aided efforts in harvesting.

Selection in the remaining two traits focused primarily on increasing the quality of the seed, which is the component of the plant that is used for food. One of the traits selected for improves a protein found in the seed, while the other is involved in determining the structure of the amylopectin, a component of the starch within the kernels. Interestingly, the researchers who studied selection for the starch enzymes have suggested that the allele chosen by the early farmers allows the starches of corn to easily form a paste. The pastiness of cornstarch allows it to form foods such as tortillas, an important contributor of calories in the Central American diet.

GREGOR MENDEL AND GENETICS

Gregor Mendel (1822–1844) is often credited with "discovering" the science of genetics. However, prior to Mendel's work in the 1900s, a number of scientists were experimenting with patterns of inheritance in agriculture and humans. For example, the study of family relationships in humans was especially important in royal families, where the documentation of marriages and births was used to establish "bloodlines" to the throne. The graphic display of these relationships is commonly called a **pedigree**, and though we usually think of them as being involved with prize horses or dogs, they are very useful in the study of human diseases that have a genetic component (see Chapter 5). Ancient Hindus used studies of pedigree to study the inheritance of traits that displayed sex-linkage, and many ancient texts provide evidence of an awareness of human heredity. In the modern era, scientists were well aware that some human diseases displayed a link with heredity. For example, color blindness has a distinct pattern of inheritance, as do diseases such as Huntington's disease and Tay-Sachs disease. However, despite the obvious importance of understanding the relationship between genetics and human disease or agriculture, scientists lacked a mechanism by which to accurately describe patterns of inheritance. Thus, they were usually limited to descriptive studies only. Very rarely was mathematics applied to inheritance. The linking of statistical methods with the study of inheritance would have to wait until the mid-nineteenth century, when a monk by the name of Gregor Mendel would introduce the first "laws" of heredity.

At first glance Gregor Mendel would appear to be an unlikely candidate for the title of "father of genetics." Born to Austrian peasants in an area now belonging to the Czech Republic, Mendel's family did not belong to the intellectual "elite" of nineteenth-century Europe. Although he demonstrated strong academic potential, especially in mathematics and natural history, his lack of personal or family wealth presented an overwhelming barrier to advancement. To overcome this, Mendel entered into the Augustinian priesthood to pursue a career in science teaching. Unfortunately, despite his interests and obvious abilities, Mendel was a poor test-taker, and he never passed his teaching certification exams. However, he did not lose his interest in the physical sciences. In fact, Mendel became even more interested in the sciences, and developed a firm belief that natural laws, including those associated with inheritance, could be explained as mathematical relationships. This form of approach, called the empirical method, was relatively new in the biological sciences, including genetics, and Mendel's place in history was established due to his ability to apply simple mathematical principles to inheritance.

For his studies, Mendel chose the garden pea, *Pisum sativum*, as his experimental organism. This choice proved to be an important one for Mendel. As in any area of the biological sciences, geneticists rely on experimental organisms for their studies. In genetics, an ideal experimental organism has three distinct characteristics. First, it must have a short generation time and have the capability of producing large numbers of offspring per generation. As we will see, the power of Mendel's work is that he linked mathematical analysis to the study of inheritance. For the most part, this is only possible if you have the ability to observe large numbers of offspring over multiple generations. Elephants or giant sequoia trees make poor experimental organisms due to long generation times. However, insects and small plants (such as peas) are ideal model organisms. The second important characteristic of a model organism is the existence of easily distinguishable traits. While modern genetics may increasingly rely on molecular or physiological traits (see Chapter 3), early researchers of inheritance needed to use organisms in which they could distinguish the offspring. Frequently, these traits focused on pigmentation differences (eye color, fruit color, etc.), but also included morphological differences (height, shape, etc.) that could also easily be identified. For Mendel, there already existed a wide range of variants for the

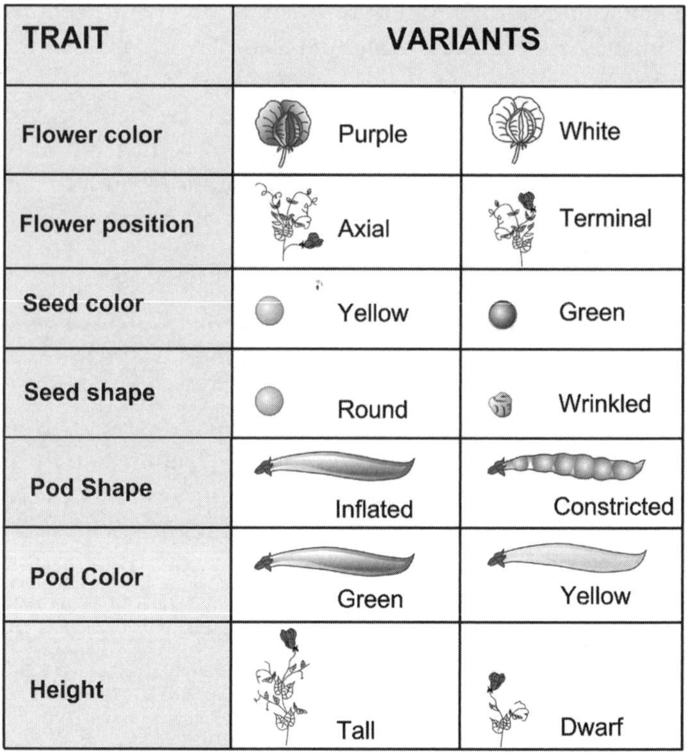

TRAIT	VARIANTS	
Flower color	Purple	White
Flower position	Axial	Terminal
Seed color	Yellow	Green
Seed shape	Round	Wrinkled
Pod Shape	Inflated	Constricted
Pod Color	Green	Yellow
Height	Tall	Dwarf

Figure 1.2 The Seven Traits that Mendel Studied in *Pisum Sativum* (*Courtesy of Ricochet Productions*).

pea plant, from which he chose seven traits as "markers" (see Figure 1.2). Finally, it must be relatively easy to grow and manipulate the ideal organism. Mendel's did not possess the resources of a modern agricultural experiment station. Instead, he was limited to a relatively small garden in which he conducted his research. Therefore, he needed an organism by which he could control breeding (which is relatively easy in plants) and grow large numbers of in a small area. Most modern experimental organisms retain this basic characteristic. Table 1.1 lists a few of the more recognized experimental organisms and what they are used for by geneticists.

Using his experimental organism, Mendel set out to examine the quantitative laws of inheritance. In his first experiment, he crossed two pea plants that possessed true-breeding variants of seed shape. These were called the round and wrinkled variants (Figure 1.2). When plants producing round seeds and plants producing wrinkled seeds are crossed, they produce a second generation of seeds, which represent

Table 1.1 Model Experimental Organisms

Species Name	Common Name	Some Research Uses
Escherichia coli	*E. coli*	Bacterial genetics
Drosophila melanogaster	Fruit fly	Development, molecular genetics, genome structure, transgenics
Mus muscalis	Mouse	Mammalian development, gene regulation
Arabidopsis thalania	Wall cress	Plant development
Caenorhabditis elegans	Roundworm	Animal development

their "offspring." When planted, these seeds yielded all plants that produce round seeds (Figure 1.3). This in itself was not surprising, as many
during Mendel's time believed that inheritance was due to a blending
of traits from one generation to another. The presence of all round
seed plants in the first generation (also called the F_1 generation) indicated that the traits were not blending together. If they were, there
should be plants of intermediate seed shape in this first generation.
However, when the round-seed-producing plants of the first generation
were crossed with one another, the plants of the second generation
(or F_2 generation) demonstrated a yield of 3/4 round-seed-producing
plants and 1/4 wrinkled-seed-producing plants (Figure 1.3).

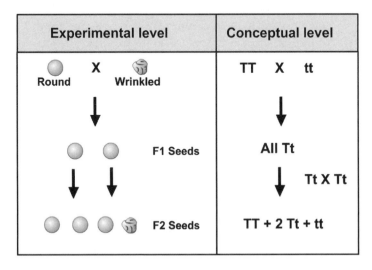

Figure 1.3 Mendel's Single-Trait Crosses (*Courtesy of Ricochet
Productions*).

Table 1.2 The Results of Mendel's Crosses

Trait	F_1 Generation	F_2 Generation	Ratio
Seed color (green or yellow)	100% yellow	6,022 yellow: 2,001 green	3.01:1
Seed shape (wrinkled or smooth)	100% smooth	5,474 smooth: 1,850 wrinkled	2.96:1
Flower color (purple or white)	100% purple	705 purple: 224 white	3.15:1
Flower position (axial or terminal)	100% axial	651 axial: 207 terminal	3.14: 1
Pod color (yellow or green)	100% green	428 green: 152 yellow	2.82:1
Pod shape (inflated or pinched)	100% inflated	882 inflated: 299 pinched	2.95:1
Plant height (tall or dwarf)	100% tall	787 tall: 277 dwarf	2.84:1

Note: See Figure 1.2 for an illustration of each trait.

These were not entirely new results. Previous work by Joseph Kölreuter, Carl Friedrich von Gärtner, and others, had already suggested that there were patterns in inheritance that could be quantified (see "What Did We Know of Genetics Before Mendel?"). However, it was Mendel who experimentally demonstrated that this was the case. Mendel's scientific and mathematical knowledge made him recognize that it was not possible to establish an empirical law of inheritance from the data derived from a single cross. So he designed and conducted six additional crosses to verify that his original observations were valid. These results are presented in Table 1.2. Notice that each of these crosses involved crossing plants with two variations of the same trait (color, height, etc.) for two generations. Two items should be evident from the data in the table. First, each of his crosses yielded approximately a 3:1 ratio in the second generation. This was consistent with his original crosses in seed shape. Second, Mendel did not confine himself to single-plant crosses, but rather replicated his experiment many times so as to increase the number of offspring. Statistically this increased the validity of his work tremendously.

⸎

WHAT DID WE KNOW OF GENETICS BEFORE MENDEL?

There were three major theories on the basis of inheritance that existed prior to Mendel. Each of these periodically gained and lost popularity over time, but remnants of each existed well into the nineteenth century. These included:

- *Pangenesis.* Pangenesis is the belief that each part of the body contributed a miniature version of itself to the offspring. For example, each finger would contribute a small version of itself, as would each ear and toe. The inheritance of nonphysical characteristics, such as behavior, presented special problems for the supporters of this theory. Variations of pangenesis were supported by a number of people, including the Greek philosopher Hippocrates, Hugo De Vries, and Charles Darwin.

- *Preformation.* The theory that a tiny preformed human, called a homunculus, resided inside of either the egg or sperm cell. This was later adapted to the idea that all parts of the adult are formed early in the development of the zygote, and simply increase in size over time. This theory was supported by seventeenth-century scientists such as Anton von Leeuwenhoek, Marcello Malpighi, and Jan Swammerdam.

- *Blending.* Under this theory, offspring represented a mixture of the hereditary material. The hereditary material was not a distinct particle, but rather a malleable substance that could be changed over the course of a few generations. This theory was supported by Joseph Kölreuter (1733–1806) and many of the other early researchers of plant hybrids.

While Gregor Mendel is considered to be the father of genetics due to his application of the empirical approach to the study of inheritance, he actually represented just one of a long lineage of individuals who were studying inheritance. Many of the earlier studies focused on the formation of hybrids in agriculturally important species of plants. Hybrids of some plants were well recognized to have improved characteristics, such as increased yield or growth rates. One of the most notable of these early geneticists was Joseph Kölreuter (1733–1806), a German botanist who worked with hybrid formation in a variety of species, including tobacco. In the nineteenth century, just prior to Mendel's time, Carl Friedrich von Gärtner (1172–1850) and Charles Naudin (1815–1899) had indicated that patterns of inheritance could be traced, that some traits exhibited dominance, and that each parent contributed to the traits of the offspring. These findings provided a firm foundation for Mendel's development of the empirical approach and it appears likely that Mendel was well aware of their work.

───────────────────────────────── ✑ ─────────────────────────────────

None of Mendel's experiments were designed to form a better pea plant, they did not increase the yield of peas, or the ability of the plant to grow. Yet these experiments are fundamentally important to an understanding of genetics. Three basic principles of inheritance were established, or supported, by Mendel's monohybrid crosses. First, his results (Figure 1.3 and Table 1.2) clearly indicated that in the first generation, one of the two variants of the trait dominated the other, or recessive, trait. This is called **dominance** (for more on the basis of dominance, see Chapter 3). Second, Mendel recognized that the pea traits he was working with represented distinct units of heredity material, and in order to get a 3:1 ratio in the second generation, each parent must possess two copies of the traits. He proposed that these two traits separate, or segregate, during reproduction, and then combine in new combinations in the offspring. This is called the **law of segregation**, or Mendel's First Law of Heredity by some. Third, the results of these single-trait crosses clearly indicated that the genetic material was not being blended. Blending would cause traits to be lost or modified in later generations, and Mendel's experiments demonstrated that recessive traits remained intact from generation to generation, although they may not be evident due to the presence of a dominant trait in the individual. This provided support for a **particulate theory of inheritance**, in which the genetic material is a physical structure passed on from generation to generation. As we will soon observe, these particles are called genes.

Mendel's work was not limited to single-trait crosses. In his writings, Mendel wondered whether the patterns he had detected in the single-trait crosses would apply to crosses involving multiple traits. He decided to test his ideas using a two-trait (or dihybrid) cross. Mendel chose the traits of seed shape and seed color for his first experiment. He recognized from his single-trait crosses (Table 1.2) that yellow seed color was dominant to green, and round seed shape was dominant to wrinkled. To facilitate analysis of his crosses, he assigned the following symbols to the traits:

A—round seed shape
a—wrinkled seed shape
B—yellow seed color
b—green seed color

This concept of assigning the dominant trait a capital letter, and the recessive variant of the trait the lower case of the same letter, is still used in many genetic systems.

In his cross, Mendel followed the same pattern as he did in the single-trait cross (Figure 1.4). First he obtained two true-breeding lines of pea plants. The first produced round and yellow seeds (AB using Mendel's symbols), while the second produced wrinkled and green seeds (symbol: ab). When these were crossed, 100 percent of the seeds produced in the first generation were round and yellow (Figure 1.4). This was not surprising, since in the single-trait crossed both round and yellow were shown to be dominant (Table 1.2). When he crossed two members of the first generation, he obtained 556 seeds in the second generation, all of which fell into one of four descriptive classes. The seeds were either round and yellow, round and green, wrinkled and yellow, or wrinkled and green (Figure 1.4). This represented all the possible combinations from the parents, but only if the traits were acting independent from one another. In other words, seed color was independent from seed shape, and these traits assorted independently in the offspring. This concept serves as the basis of Mendel's **law of independent assortment**, or Second Law of Heredity.

An examination of the number of seeds in each class indicates that the offspring are not distributed evenly among the classes. Mendel recognized that there was a distinct ratio present in the second generation seeds. The seeds with two dominant traits, round and yellow, were represented 9/16 of the time. There were two classes of seeds that possessed only one of the dominant traits: round and green and wrinkled and yellow. Each of these was present in approximately 3/16 of the offspring. In comparison, there were rel-

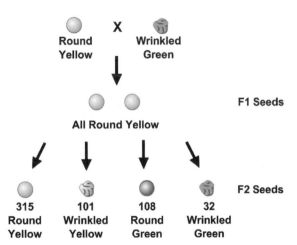

Figure 1.4 Mendel's Experiments with Two-Trait (Dihybrid Crosses) (*Courtesy of Ricochet Productions*).

atively few seeds displaying the two recessive traits (wrinkled and green). These only occurred 1/16 of the time.

By the laws of **probability**, the probability of getting independent events is equal to the sum of their individual probabilities. If we dissect a two-trait cross into two one-trait crosses, then the probability of getting a recessive trait in the second generation is 1 in 4 (1/4). Thus the probability of getting two recessive traits is equal to the product of their individual probabilities (1/4 for seed shape X 1/4 for seed color), or 1/16. This is exactly what Mendel observed in his two-trait cross. Once again, Mendel was intelligent enough to recognize that a single cross was not a verification of his ideas, so he designed additional crosses under the same parameters. All of these yielded similar results, providing an additional verification of his law of independent assortment.

Mendel's paper, "Experiments in Plant Hybridization" (1865) was published in the *Brünn Scientific Society Proceedings*, a scientific journal of the mid-nineteenth century. Unfortunately, Mendel's work was not widely received by the scientific community, although the paper itself is easy to follow by modern standards. Many books suggest that this was because Mendel published in a lesser-read journal. This does not appear to be completely accurate and other historians suggest that the reason for the lack of attention to Mendel's work is due to the fact that he was ahead of his time. Regardless of the reason, Mendel's work had almost no influence on the scientific community of the time, and quickly fell into obscurity. However, 35 years later, the scientific world caught up to Mendel.

IMPORTANT GENETIC TERMS

The genetic terminology of Mendel's time was rather limited. Mendel introduced the concepts of dominance and recessive, as well as some of the symbolism, namely the use of upper- and lowercase letters that is typically associated with the identification of variants of traits. However, before proceeding with any discussion of genes and gene function, it is important to provide an overview of some important terms.

Mendel dealt almost entirely with observing the physical characteristics of the organism. This is called the **phenotype**. However, he did recognize that the expression of a trait was actually due to the combination of traits within the organism. The instructions for producing the trait are called the **genotype** of the organism.

Second, the instructions for producing a trait are contained within the DNA of the organism. The material in Chapter 2 will cover the science of how we know this to be true. Within the DNA are regions of information called *genes*. The term gene was first introduced by Wilhelm Johannsen (1857–1927) in 1909. Relating back to Mendel, flower color and seed shape are both determined by genes. Each gene may have a number of variants, which geneticists call **alleles**. An allele is due to a minor change in the genetic material (see Chapter 3). The genotype is the combination of alleles in an organism.

Genes are located on *chromosomes* (this is explained more in the next chapter) at specific locations, called a **locus**. Think of chromosomes as filing cabinets for genes. Each species has a specific number of filing

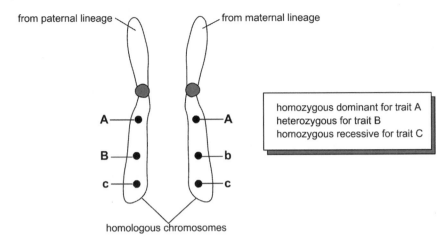

from paternal lineage

from maternal lineage

A — A
B — b
c — c

homozygous dominant for trait A
heterozygous for trait B
homozygous recessive for trait C

homologous chromosomes

Figure 1.5 Genetic Terminology of the Chromosome (*Courtesy of Ricochet Productions*).

cabinets (chromosomes), and the genes are arranged in a specific order (loci) within the cabinet. However, for any gene there may be variations in the information that it contains (alleles).

Mendel determined that organisms such as peas have two copies of each trait. While this is not true for every organism, it does apply to animals and plants (and many others). We get one set of traits from our father, and one set from our mother. Since these traits are determined by genes and genes are located on chromosomes, we can say that we get one set of chromosomes from each parent. The information on each chromosome in the pair is arranged in the same way (usually), but since genes may have different alleles, the chromosomes in a pair are rarely ever truly identical. We call these similar chromosomes **homologous**. Since we have a pair of each chromosome, we are called **diploid** organisms.

For any trait, a diploid individual can have either two identical alleles for the trait, or two different alleles. If you have two of the same allele for a given trait, you are said to be **homozygous** for that trait. If you have two different alleles, you are called **heterozygous**. A *homozygous recessive* individual has two identical recessive alleles for the trait being studied, and a *homozygous dominant* individual has two dominant alleles. See Figure 1.5 for an overview of these terms.

It is often easier to understand genetic relationships using diagrams. One of the most common diagrams that is used in the study of Mendelian genetics is the Punnett square. It was originally designed by the English

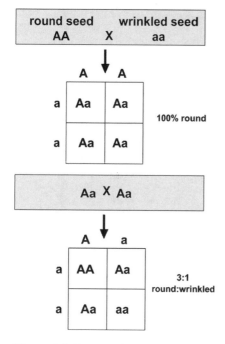

round seed   wrinkled seed
   AA   X   aa

	A	A
a	Aa	Aa
a	Aa	Aa

100% round

Aa **X** Aa

	A	a
a	AA	Aa
a	Aa	aa

3:1
round:wrinkled

Figure 1.6 Punnett Square Diagram of a Single-Trait Cross (*Courtesy of Ricochet Productions*).

geneticist Reginald Punnett (1875–1967) in 1909 to explain basic inheritance in his introductory genetics textbook. The concept of a Punnett square is relatively simple (Figure 1.6). The symbols on the outside of the box represent the allele combinations in the gametes for the traits being studied. The symbols within the box represent the possible offspring (zygotes) from the combination of the gametes. A Punnett square is useful in illustrating the potential offspring of a given cross when the inheritance patterns of one or two traits are being examined. For example, Figure 1.6 presents a Punnett square diagram of Mendel's single-trait cross with seed shape (from Figure 1.3). However, the diagram becomes cumbersome for more than three traits, so most geneticists rely upon other means, including statistics, to predict the outcome of complex crosses.

THOMAS HUNT MORGAN AND THE FLY LAB

In 1900 three botanists, Hugo de Vries (1848–1935), Erich Tschermak (1871–1962), and Carl Correns (1864–1933), independently published results of their research that effectively "rediscovered" Mendel's work. de Vries' research examined crosses in 11 different species of plants, in which he confirmed Mendel's idea of dominant and recessive traits and the ability to predict the ratio of traits in the offspring. Like Mendel, Tschermak studied peas (*Pisum sativum*) and observed 3:1 ratios in the second generation offspring of single-trait crosses. Correns studied the formation of hybrids in both peas and corn, and once again obtained results similar to that found in Mendel's studies. Some historians debate what each of these individuals knew of Mendel prior to publishing their work. However, what is truly important is the fact that Mendelian inheritance was back in the forefront of genetic studies, and the result was a decade-long flurry of discoveries and discussions on the properties of inheritance. One of the leading contributors to genetics during this time was the embryologist turned geneticist, Thomas Hunt Morgan (1866–1945).

Anyone who studies genetics quickly becomes familiar with a species called *Drosophila melanogaster,* commonly called the fruit fly (or vinegar fly). The fruit fly is not considered to be a pest species of agriculture (except for some cases in the wine industry) and is not known to be a **vector** for human disease. Therefore, to most people it is unclear as to why geneticists sometimes appear to be preoccupied with this species of insect. Yet we know more about this species of animal than any other, including ourselves. The reason why is partly due to history, and partly due to the fact that this organism is an ideal model species for the studies of genetics.

As presented in the entry on Mendel and the laws of genetics earlier in this chapter, Mendel chose the pea plant for his studies of inheritance for several reasons. First, the pea plant was relatively easy to grow and could easily generate true-breeding lines for specific traits. Second, it produced large numbers of offspring (peas). And finally, it has easily identifiable traits (flower color, height, etc.). In most cases, these are the general criteria for many of the model organisms used by genetic researchers. For many geneticists, the organism itself is not the focus of the study, but rather what the organism can tell us about inheritance and the genes responsible for traits. Such is also the case for the fruit fly, *Drosophila melanogaster.*

The study of fruit fly genetics began in earnest in the lab of Thomas Hunt Morgan (1866–1945), at Columbia University in the early twentieth century. Morgan was an embryologist by training, but conducted research in a wide range of subjects, including studies of how mutations influenced populations. Through correspondence with other scientists, Morgan developed an interest in using fruit flies over other lab organisms, such as rats and chickens. Fruit flies were relatively easy to grow, had a short generation time, and had plenty of vigor. They were also easy to manipulate in the lab. In the early twentieth century, the researchers in Morgan's lab made several important contributions to the young science of genetics.

As noted in the earlier entry entitled "Corn and Wheat: Birth of Civilization and Genetics," the concept of sex-linkage was not unknown in the study of inheritance. Human diseases such as color blindness are known to be inherited in a sex-specific manner. It was also known that while males and females have basically the same number and types of chromosomes, they do differ in the composition of one set, the sex chromosomes. Thus, it makes sense that **sex-linked traits** may somehow be associated with these chromosomes. It was Morgan's work that provided some of the first experimental evidence that this was the case, and eventually led to the confirmation of the chromosomal theory of inheritance.

One of Morgan's prime research interests was in the study of mutations. For almost 2 years he raised *Drosophila* under a variety of conditions (including exposure to X-rays and radium) in an attempt to detect new mutations. In 1909 he discovered a single white-eyed male in one of the samples. The normal eye color (also called **wild-type**) for *Drosophila melanogaster* is red. Morgan then designed a series of crosses to determine the basis of his white-eye mutation (Figure 1.7).

From the illustration you can observe that the first generation of flies from Morgan's cross all possessed red eyes. However, when the first generation flies were allowed to interbreed, the second generation offspring yielded some interesting results. All of the females possessed red eyes, but only 1/2 of the males. The other 1/2 of the males had white eyes. The reason for that can be determined by examining the chromosomes under each of the flies. Male flies (like humans) have an X and Y chromosome, while females have two X chromosomes. Since the male has a single X, any trait on the X acts as if it is dominant, meaning that it will be expressed in the males. However, females have two X chromosomes, and thus must possess two copies of a recessive trait for it to be expressed.

Notice that the white-eye mutation (X^w) is being passed from the male parent to the females of the first generation, making them *heterozygous* for the white-eye trait. The males of the second generation always get their single X chromosome from the female, and thus as normal since the females have two normal X chromosomes (X^{w+}). The "+" sign indicates that this chromosome is normal (wild-type) for the white trait. However, in the second generation, the females will give 1/2 of their male offspring an X^{w+} chromosome, and 1/2 an X^w chromosome, thus producing the phenotypes of the second generation males. Since all females in the second generation received one chromosome from the wild-type male parent, they are all phenotypically normal.

In these experiments, Morgan and his colleagues were successful in identifying that the information for a trait is located on a chromosome. However, at first Morgan didn't totally comprehend the importance of what they had done. In the late nineteenth century, before Morgan began his work with *Drosophila*, there was a developing interest in understanding what represented the genetic unit of inheritance. Carl

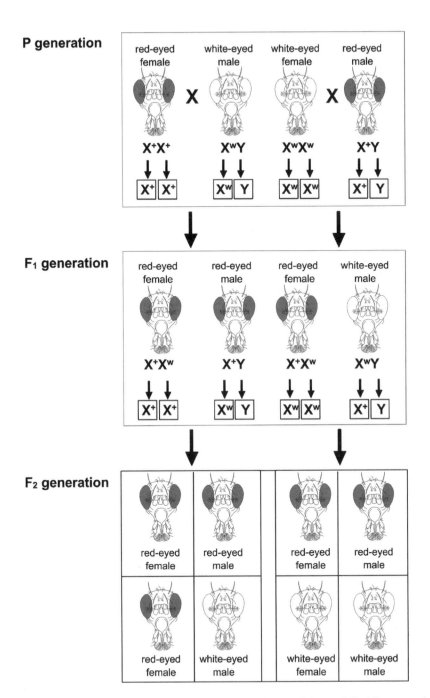

Figure 1.7 Morgan's Experiments with White-eyed *Drosophila* (*Courtesy of Ricochet Productions*).

Nägeli (1817–1891) and August Weismann (1834–1914) recognized that some compound in the nucleus was responsible for transmitting traits from generation to generation. Other scientists suggested that it was the chromosomes, first identified in the late nineteenth century that housed the genetic material. Around 1902, Walter Sutton (1877–1916) and Theodore Boveri (1862–1915) were independently studying the patterns of chromosomes during **meiosis** when they recognized that the chromosomes were following a Mendelian pattern of inheritance. From these observations they developed the chromosomal theory of inheritance, which has the following main components.

- The genetic material is contained in the chromosomes.
- Chromosomes are replicated and passed on from generation to generation.
- Most eukaryotic cells contain two copies of each chromosome. These chromosomes segregate to form **haploid** gametes during meiosis (in animals).
- Each parent contributes one set of chromosomes to its offspring via the gametes.

The chromosomal theory was correct, but it lacked a proof. What was needed was an experimental system that placed a trait on a chromosome, thus allowing the chromosome to be tracked from generation to generation. The initial proof of the chromosomal theory was provided by Morgan and his white-eyed fruit flies. Notice from Figure 1.7 that you can track the movement of the white-eye trait from generation to generation, and that the trait is located on the X chromosome. Additional proof of the chromosomal theory was provided by Calvin Bridges (1889–1938), a researcher in Morgan's lab who assisted with a large amount of Morgan's work after 1910. Together Morgan and Bridges had succeeded in placing traits on chromosomes, and effectively linked studies of cell division with that of genetics.

By 1911 the Morgan lab had identified a number of mutations in *Drosophila* that could be easily identified. In addition to the white-eye mutations, there were variations in wing shape and body color, as well as additional mutations for eye color. What is interesting is that the lab had identified six mutations that were all known to be sex-linked and located on the X chromosome. One of Morgan's research team, an undergraduate named Alfred Sturtevant (1891–1970) realized that if all of the genes were located on the same chromosome, then it should be possible to determine a relative map of the chromosome. However, before examining Sturtevant's work, we need to first provide some background information on what happens to chromosomes during meiosis.

During meiosis, homologous chromosomes exchange genetic information by a process called *crossing over*. This occurs at the molecular level of the chromosome, but what it effectively does is to shuffle the genes between chromosomes. The genes themselves remain at the same location on the chromosome, but the genetic material from one chromosome is swapped with the exact same genetic material from another chromosome. Imagine having two identical volumes of an encyclopedia, except that one is written in red ink, while the other is written in black ink. If these were chromosomes, crossing over would take pages 200–215 of one volume and replace it with pages 200–215 of the other volume. The content would remain basically the same in both volumes. The process is similar in crossing over, except that sometimes we swap alleles between the two chromosomes. Recall that alleles are variations of a gene, so although the chromosomes remain relatively similar after crossing over, they may contain new combinations of alleles.

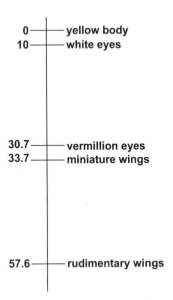

Figure 1.8 The First Genetic Map (*Courtesy of Ricochet Productions*).

The degree to which crossing over will occur is dependent upon many factors, primary of which is the distance between the two genes. The closer the genes are on the chromosome, the less likely crossing over will occur. We can use this information to establish relative distances between genes on the same chromosome. This is called the map distance and is represented by the following formula:

$$\text{Map distance} = \frac{\text{Numbers of recombinant offspring}}{\text{Total number of offspring}}$$

The term **recombinant** is used to describe those gene combinations that are present in the offspring that represent a combination that is not present in one of the two parents. Gene combinations that are the same as those found in one of the parents are called **nonrecombinants**. In the formula above, the total number of offspring represents the sum of the recombinant and nonrecombinants in the cross.

Sturtevant recognized that this relationship existed for the sex-linked traits being studied in Morgan's lab. In 1911, using the available information from the multitude of crosses being conducted in the lab, Sturtevant was able to construct the genetic map presented in Figure 1.8. Notice from the diagram that several of the traits are located close together (yellow body and white eyes), while others are located considerable distances from each other on the chromosome. The numbers represent the percent of recombinant offspring that were observed in crosses involving the two traits. For example, in crosses that used white eyes and vermilion wings, 20.7 percent of the offspring were recombinants. It is important to note that these are relative distances, not actual locations on the chromosome. Modern techniques have the ability to pinpoint the physical location of genes on a chromosome in a variety of species. However, the construction of Sturtevant's map is considered by many to be an important milestone in the study of genetics.

IS INHERITANCE AS SIMPLE AS MENDEL SUGGESTED?

A quick examination of the natural world appears to support the application of Mendel's laws to inheritance. In general, offspring resemble their parents and traits can be tracked from generation to generation. However, a closer look will reveal that in many cases Mendel's laws do not apply. A cross between a red and white flower may yield pink offspring, suggesting that blending is occurring. Certain traits seem to be more prevalent in one sex than the other. Furthermore, for most traits, including many of those in humans, we are not simply a choice between two variants of a gene. A simple look at eye color in humans reveals that there are many differences in the population.

The question may be asked as to whether Mendel was correct in his assessment of inheritance. The answer is both yes and no. Mendel was absolutely correct in his support of the particulate theory of inheritance. Parents pass distinct particles (the genes on chromosomes) to their offspring. Of even greater significance was Mendel's application of mathematical principles to the study of heredity. The use of empirical analysis is a cornerstone of genetic studies, you would be hard pressed to find a geneticist who does not use mathematical laws in their research. But the natural world is much more complex than Mendel described, and it was not long after Mendel's work was rediscovered before early geneticists began to describe patterns of inheritance that deviated from the mathematical Mendel's laws. The next few sections in this chapter examine a few variations in Mendelian principles. It should be noted that this is not a complete list, there are entire textbooks dedicated to the study of non-Mendelian traits and inheritance. The items below have been chosen to give a sampling of variations from the laws that Mendel described.

Incomplete Dominance

Around 1905, the German scientist Carl Correns (one of the individuals who rediscovered Mendel's laws) was examining the basis of inheritance in the four o'clock plant (*Mirabilis jalapa*) when he noticed what at first appeared to be evidence of blending. In his experimental system, Correns crossed red-flowered plants with white-flowered plants. If the trait was following basic Mendelian principles, you would expect to obtain a first generation of all one-color plants. This would indicate the dominant allele. When Correns performed his crosses, all of the offspring were pink. While red may be considered to be dominant, there was no enough expression of the red pigment to mask the white pigment. This is called **incomplete dominance**.

When Correns crossed two pink plants from this generation, their offspring were 1/4 red, 1/4 white, and 1/2 pink. Thus the white and red alleles were not "lost" in the first cross, and must have of remained intact in order to be expressed in the offspring of the second cross. The question can be asked as to why the phenotype of the first generation differ from that observed in Mendel's single-trait cross (Figure 1.3). The answer involves a quick examination of **gene expression**. Most genes contain instructions for making proteins. In turn, these proteins produce the phenotype of the individual. In Mendel's crosses, the dominant gene produced enough protein product to completely mask the protein product of the recessive allele. The dominance of the first allele was complete. This is not the case for the flower color of four o'clock plants. In this case the protein does not completely mask the influence of the recessive allele, and thus a blending of protein function occurs.

Incomplete dominance may look like the perfect proof for the pre-Mendel theory of blending (see "What Did We Know of Genetics Before Mendel?"). However, the original concept of blending was proposed to describe blending of the genetic material. The experiments with the four o'clock plants is a blending of phenotypes due to an inability of the protein being produced by the red alleles to mask the white traits. Since the white trait is "recovered" in the second generation, the genetic material was not blended. Instead the interaction of the proteins that produce flower color produced a blended phenotype. Thus, incomplete dominance can be considered to be a case of phenotypic blending, not genotypic blending. Other organisms display incomplete dominance,

Table 1.3 Human Blood Groups

Blood Type	Possible Genotypes	Forms Antibodies Against
O	ii	A, B, or AB
A	$I^A I^A$, $I^A i$	B
B	$I^B I^B$, $I^B i$	A
AB	$I^A I^B$	None

such as the color of the fruit in eggplants and spotting patterns in horses.

Multiple-Allele Systems

For most of the crosses we have examined in this chapter, the phenotype has been limited by the fact that we were dealing with a two-allele system. However, for the traits of many species there are more than two alleles in the population. A gene that has multiple different alleles in a population is called **polymorphic** (many forms). An excellent example is human eye color, where there are at least three alleles in the population that interact to produce the wide range of eye colors we observe in our species. The analysis of multiple-allele systems can become complicated since the concept of dominant and recessive does not completely hold true. Sometimes, two alleles can be dominant at the same time. This is called **codominance**. In other cases there is more of a hierarchy, where one allele is dominant over another, but recessive to a third. This occurs in the coat color of rabbits where there are four alleles interacting.

An example of a multiple-allele system in humans is the human blood groups. Blood type is determined by genes containing instructions for proteins in the membranes of red blood cells. These proteins are called *antigens*, because they are the structures that interact with the **antibodies** produced by B cells of the immune system. If a cell does not have the correct antigen it will be identified by antibodies as foreign and targeted for destruction by the defense mechanisms of the immune system.

There are three different antigen alleles that are possible in humans. These alleles produce slight variations in the structure of the cell surface proteins. Two of these alleles are dominant (I^A and I^B). In individuals that possess both an I^A and I^B allele, the alleles function in a *codominant* manner, with two different types of antigens (A and B) being placed on the cell surface. There is also a third allele (i) that is recessive to both I^A and I^B. I^A and I^B are both completely dominant over i. These alleles form the basis of the human blood groups (Table 1.3).

A person who possesses the I^A allele will produce antibodies against I^B proteins on the surface of foreign blood cells. Individuals with I^B alleles will produce antibodies against red blood cells with I^A proteins. However, neither will produce antibodies against cells that are homozygous for *i*. This is because the *i* allele produces a cell surface protein that it not recognized by antibodies. Individuals who have both I^A and I^B alleles will not produce antibodies, since their cells contain both types of cell surface proteins. This information is important in blood transfusions involving whole blood (red blood cells). If a person with type A blood (see Table 1.3) receives type B blood, they will produce antibodies against the blood, which will cause clotting and severe medical problems. Notice from Table 1.3 that type O blood will produce antibodies against anything but O blood, but type AB blood will not produce antibodies against any other blood types. For this reason, people with AB blood are often called universal acceptors, while people with O blood are called universal donors. This is just one example of a multiple-allele system.

Gene Interactions

At one time scientists predicted that the final analysis of human **genome** would uncover between 100,000 and 150,000 genes. They believed that only a large number of genes could possibly produce as complex an organism as a *Homo sapien*. Over the years that number has dropped to 50,000, then 35,000. With the completion of the **Human Genome Project**, it is now believed that there are around 22,000 genes in our genome. Humans are not less complex than we originally thought. Instead, our complexity is based on gene interactions.

A nice example of how genes can interact to produce new phenotypes is coat color in Labrador retrievers. Anyone familiar with this breed of dog knows that there are three phenotypes: black, brown (chocolate), and yellow. However, these three phenotypes are determined by only two genes. The first of these genes is responsible for producing a black or brown coat color. The dominant allele is black (B) and the recessive allele is brown (b). This gene is expressed in the melanocyte cells, which are responsible for producing the pigments associated with coat color. A second gene determines whether the pigment is deposited inside of the hair cells. The dominant form of this gene (E) allows the pigment to be deposited, but the recessive form (e) does not. If neither a black or brown color is deposited in the hair cell, the hair takes on a yellow color. A dog that is homozygous for this gene will not display the phenotype of the first gene. This is called **epistatis**, which means that the expression of one gene masks the expression of a second gene.

Figure 1.9 demonstrates an example of a two-trait cross between two black retrievers that are heterozygous for both traits (BbEe). Notice how the traits interact with one another and that yellow retrievers possess the genetic information to be a black or brown dog. These coat color alleles are simply not being expressed in the hair cells of the yellow retrievers. Recall from the Mendelian two-trait cross (Figure 1.4), the crossing of two heterozygous individuals produced a 9:3:3:1 phenotypic ratio in the offspring. In the cross involving the retrievers, the offspring display a 9:3:4 phenotypic ratio (black: brown: yellow). This is a characteristic of traits that involve gene interaction, and there are many of them in the natural world. Frequently these types of interactions are evident in the enzymes of metabolic pathways in a species.

BbEe X BbEe
black   black

	BE	Be	bE	be
BE	BBEE black	BBEe black	BbEE black	BbEe black
Be	BBEe black	BBee yellow	BbEe black	Bbee yellow
bE	BbEE black	BbEe black	bbEE brown	bbEe brown
be	BbEe black	Bbee yellow	bbEe brown	bbee yellow

Figure 1.9 Gene Interactions Determining Coat Color of Labrador Retrievers (*Courtesy of Ricochet Productions*).

2

DNA AS THE HEREDITARY MOLECULE

Before proceeding to the discussion of how genes operate and interact to produce a living organism, it is important to recognize the contributions of the scientists who determined that DNA is the hereditary material. In today's world, discussions of DNA-related issues, such as cloning and **transgenic** organisms, are an almost nightly occurrence on the evening news. It is sometimes hard to imagine that just over 50 years ago chemists and biologists were struggling to determine the structure of DNA, and that it has only been a relatively short period of time since the scientific community first recognized that DNA is the hereditary material of all living organisms.

In this chapter we will examine a few of the key discoveries in the history of DNA. This is not simply just another lesson in history, but rather a review of some of the classical scientific experiments of all time. In the study of science, the **scientific method** represents the central ideology under which science functions. In the scientific method, scientists take observations of the natural world and then try to determine the natural laws or conditions that are responsible for the observation. This is done by experimentation, the identifying characteristic of scientific discovery. Before designing an experiment, a researcher must develop a hypothesis, or tentative explanation, for the observation to be made. Hypotheses must be fairly specific, and typically address one variable or factor that may be contributing to the factor. This is important because experiments are designed to test standardized conditions to the influence of a single variable. Broad-based experiments often yield inconclusive results. Furthermore, scientists are not surprised when experiments fail to support their initial hypothesis. Instead, they use the information from the experiment to develop new hypotheses and experiments. Frequently it may take multiple rounds of hypotheses and experiments to study a single observation.

Each of the experiments outlined in this chapter represent a model application of the scientific method. The researchers involved in the study asked specific questions, and then designed an experiment that limited the variables involved so that they could test their specific hypothesis.

GRIFFITH AND THE TRANSFORMATION OF DNA

Following the rediscovery of Mendelian principles in the 1900s, early geneticists were actively determining many of the basic principles of inheritance. But one question loomed large in the scientific world, what was the genetic material? The road to the discovery of DNA as the genetic material is frequently credited as beginning with the work of Frederick Griffith (1877–1941), an English bacteriologist. Griffith's interests were in developing a means of assisting the immune system in fighting off forms of pneumonia. The bacteria that causes this disease is *Streptococcus pneumonia.* As is the case with many **prokaryotic** cells, the bacteria that Griffith worked with possessed a **polysaccharide** (sugar) covering around the outside of the cell. This type of coating is frequently used by disease-causing bacteria to avoid the immune responses of the host. Strains of *S. pneumonia* that possessed this coating were disease-causing or virulent. Furthermore, when they were placed on a growth medium, they formed a smooth colony (or plaque). For this reason the strain was often referred to simply as the smooth (S) strain. In addition to the smooth strain, the lab that Griffith worked in had isolated a second, non-virulent, strain. This lacked the polysaccharide coat, and when plated on media formed rough shaped colonies. Thus it was called rough (or R). If you are trying to develop a vaccine, one goal is frequently to use a nonvirulent strain to develop immune protection against a virulent strain.

Griffith used mice as an experimental system. He developed several treatments involving combinations of S and R strains (see Table 2.1). A mouse injected with S bacteria died since its immune system was unable to fight off the infection. This was evident by the presence of live S strain bacteria in the blood of the dead mouse. The exact opposite was true of R strain bacteria. Not only did the mouse live, but its immune system had removed bacteria from the blood system (Table 2.1). The third experiment involved heat-killing the S bacteria. As is evident from the data, this effectively destroyed the ability of the bacteria to cause an infection, and the mouse lived.

Table 2.1 Griffith's Experiments with *S. pneumonia*

Injected Strain	Condition of Mouse	Recovered Strain
Live S	Died	Live S
Live R	Lived	None
Heat-killed S	Lived	None
Live R + heat-killed S	Died	Live S

Note: The S denotes the smooth (virulent) strain, while the R denotes the rough (nonvirulent) strain of the bacteria.

The crucial test occurred when Griffith injected heat-killed S bacteria along with live R strain into the same mouse. Individually, both of these treatments had not resulted in mortality, but together (Table 2.1) they caused death in the mouse. When blood from the dead mouse was analyzed, Griffith found evidence of live S strain bacteria. This could only be possible if something from the dead S strain was altering the physiology of the R strain to cause it to be virulent. This "substance" had to contain the instructions for the manufacture of the polysaccharide coat, thus it was very likely that what was being transferred was genetic material. Griffith called this substance a "transforming principle," but he was not successful in determining exactly what it was.

Transformation is an important aspect of bacterial genetics. Not only can bacteria pick up genetic material from its environment, a process called *gene transfer*, but scientists can manipulate bacteria in the lab by transforming them. This will be covered again in Chapter 4. However, it is important to note that following Griffith's work, other scientists began to recognize that an isolation of the transforming principle would provide an insight into the nature of the genetic material.

AVERY AND THE TRANSFORMING PRINCIPLE

Following Griffith's identification of transformation, a number of researchers began to look for a method of determining the identity of the transforming substance. There were three prime candidates for the genetic material. The first was proteins, which were known to be involved with almost all aspects of cellular physiology, from metabolism to reproduction. The second group consisted of the nucleic acids, deoxyribonucleic acid (DNA) and ribonucleic acid (RNA). The exact role of these compounds in the cell was still not clear, although it was known the DNA was found in the nucleus, while RNA was found throughout the cell.

Table 2.2 The Avery Experiments to Identify
the Transforming Substance

Treated With	Removes	Transformation of Type R?
Proteases	Proteins	Yes
Rnase	RNA	Yes
Dnase	DNA	No

Note: In each experiment, cellular material from dead
type S (virulent) bacteria were treated with an enzyme
to remove a specific molecule. The resulting extract
was then administered to type R bacteria to determine
if the type R were transformed to type S.

A team of researchers at the Rockefeller Institute, a leading research
establishment, led by Oswald Avery (1877–1955), recognized that
by using biochemical purification techniques it may be possible to
separate the candidate materials. Once isolated, Griffith's experiment
could be repeated using specific extracts, and the genetic material
identified. Unfortunately, the isolated techniques of the late 1930s and
early 1940s were not perfect, and it took a considerable amount of time
for Avery's team to obtain pure samples. However, by 1944 they had
designed an experimental system to test the identity of the transforming
substance.

Their experimental system was relatively simple by modern standards
(Table 2.2). Although simple, it is widely regarded as a model example
of the scientific method. Avery and his colleagues developed an experimental system that used a series of experiments that were identical, with
the exception of one factor. The procedure isolates the influence of a
single variable to the phenomena being studied, which in this case was
the nature of Griffith's transforming principle.

In their experimental design the researchers first obtained cultures
of heat-killed type S (virulent) bacteria. This culture possesses the ability
to transform type R bacteria, as described previously by Griffith. Avery
and his colleagues then took samples of the culture and treated it with
digestive enzymes. Each of these enzymes was capable of breaking down
one specific class of molecules in the culture. To eliminate the influence
of proteins, the culture was treated with general proteases. The treated
culture was then administered to a second culture containing type R
bacteria. Since the treated culture retained the ability to transform type
R into type S (see Table 2.2), then the missing proteins were not acting

as the transforming agents. A similar experiment in which the type S culture was treated with RNase to remove RNA yielded the same results. Thus, RNA was not the transforming agent. However, when the type S culture was treated with DNase, the culture lost its transforming properties. In the experiments with protease and RNase treatments, the DNA was unaffected. When the results of the DNase treatment were coupled with the RNase and protease treatments, it was apparent that DNA was acting as the transforming substance.

Avery and his colleagues believed that they had conclusively proven that DNA was the genetic material, but the scientific community was not entirely convinced. Biochemical purification in the 1940s was not 100 percent totally reliable, thus there was the possibility that some DNA was still present in the samples. Furthermore, at the time many scientists firmly believed that proteins were the hereditary material, and that while DNA may be acting as a transforming factor, that did not necessarily prove that it was also the genetic material. However, 8 years after Avery's experiments, conclusive evidence was delivered to the scientific world that DNA was the hereditary material.

DNA OR PROTEIN? THE HERSHEY-CHASE EXPERIMENTS

In 1952 Alfred Hershey (1908–1997) and Martha Chase (1928–2003), of Cold Spring Harbor Laboratories, provided the final experimental evidence needed to convince the scientific community that DNA is the genetic material. As was the case with Avery's work, the design of the Hershey–Chase experiments is considered by many scientists to be a model example of the scientific method.

In their studies Hershey and Chase recognized that the ideal experimental system with which to study whether DNA or proteins is the genetic material was a **virus**. Viruses are infectious systems that are much smaller than cells. Most viruses primarily consist of proteins and nucleic acids, although some also contain lipids. Viruses lack the ability to reproduce on their own, instead they must infect a host cell and pirate its metabolic machinery into manufacturing new viruses. There are many types of viruses (HIV, cold virus, etc.), but in general viruses are very specific in the types of cells that they will infect. The type of virus that Hershey and Chase worked with is called T2, and it belongs to a class of viruses called the **bacteriophages**. Bacteriophages (also called *phages*) are viruses that infect bacterial cells. The bacteriophage T2's host cell is the bacterial *Escherichia coli* (or *E. coli*). *E. coli* is a popular bacteria among scientists, as it is easy to culture in the lab and much is known about its physiology and genetics.

All viruses infect cells by first attaching to the cell wall of the host cell, and then injecting genetic material into the **cytoplasm** of the cell. Scientists already knew that following infection of the host cell, new viruses were manufactured that were genetically the same as the "parent" virus, thus it was evident that the virus was providing genetic material for the next generation. Since viruses are primarily protein and nucleic acids, one of these must be the genetic material. Hershey and Chase developed an experimental system in which the proteins were labeled with radioactive elements, or **isotopes**. These isotopes would be used as markers for the proteins and nucleic acids in the bacteriophage. Isotopes of an element behave metabolically just like a nonradioactive atom of the same element, so the use of isotopes should not influence the ability of the virus to infect the *E. coli* cell. What Hershey and Chase needed was one element that was unique to protein, and a second element that was unique to the nucleic acids (specifically DNA). They used a radioactive form of sulfur called ^{35}S. Sulfur is found only in proteins, and not in DNA. As a marker for DNA they choose radioactive ^{32}P, since phosphorous is only found in DNA.

The experimental design of the Hershey–Chase experiment was simple. One batch of viruses were grown in the presence of ^{35}S, while a second batch was grown in the presence of ^{32}P. These radioactive T2 viruses were then allowed to infect *E. coli* cells. After allowing the infection to proceed for a short period of time, the researchers removed the radioactive protein shells of the virus by agitating the solution in a blender. The infected *E. coli* cells were then allowed to grow on a nutrient media until new viruses were formed within the cells.

The researchers then examined what radioactive element was present within the infected *E. coli* cells and newly formed bacteriophages. Very little radioactive sulfur (^{35}S) was found, most of it had been removed by the agitation action of the blender. Thus it was apparent that the proteins were not acting as the genetic material, since it was not entering the cell to direct the formation of new T2 bacteriophages. However, the researchers could detect significant amounts of radioactive phosphorous (^{32}P) in the cells, supporting the idea that the bacteriophage was transferring its instructions into the cell as DNA. When coupled to the experiments of Avery and his colleagues, the evidence provided by the Hershey–Chase experiments conclusively demonstrated that DNA, not proteins, is the genetic material. Now that the nature of the genetic material had been identified, it was time to determine its structure.

THE STRUCTURE OF DNA

With the possible exception of a few of the larger proteins, DNA represents one of the most structurally complex molecules in biological systems. In today's world, we all readily recognize the double helix as the representative molecule of the biological sciences. However, less than a century ago, scientists were struggling with understanding the structure of this all-important molecule. The discovery of DNA structure is not the story of a single breakthrough experiment by a lucky researcher, rather it represents a series of small discoveries that each provided a hint as to the structure of the genetic material. Only after a significant number of these clues had been determined was it possible to put together a DNA model.

Chargaff and Complementation

The building blocks of DNA are the **nucleotides**. Each nucleotide consists of three subunits: a five-carbon sugar, a phosphate group, and one of four nitrogenous bases (see Figure 2.1). The chemical composition of the nucleotide was worked out in the late nineteenth and the early twentieth centuries by chemists such as Albrecht Kossel (1853–1927) and Phoebus Levene (1869–1940). However, it was not until around 1948 that a biochemist named Erwin Chargaff (1905–2002) discovered an interesting relationship of the nucleotides within a sample of DNA.

As a biochemist Chargaff was very familiar with the chemical procedures for isolating DNA from tissues. Since the only item in a nucleotide that varies is the choice of nitrogenous base, then it must be these items that contain the genetic information. Chargaff understood that it was possible to identify the percent of each nitrogenous base by subjecting the DNA to a series of reactions that would purify and isolate the bases. Starting with tissues from seven different species, Chargaff purified the DNA and then exposed it to a strong chemical base in order to break the bonds holding the nucleotide together.

Once the bases had been isolated, Chargaff set out to determine the proportion of each base in the sample using a *spectrophotometer*. A spectrophotometer is an instrument that has the ability to generate light at specific wavelengths. Once generated, the light is passed through the sample to a detector. However, only a fraction of the original wavelength will be transmitted through the sample to the detector, the rest will be absorbed by the sample. All chemical compounds have specific wavelengths at which they exhibit maximum absorbance. This is typically based on the chemical structure of the compound. These wavelengths

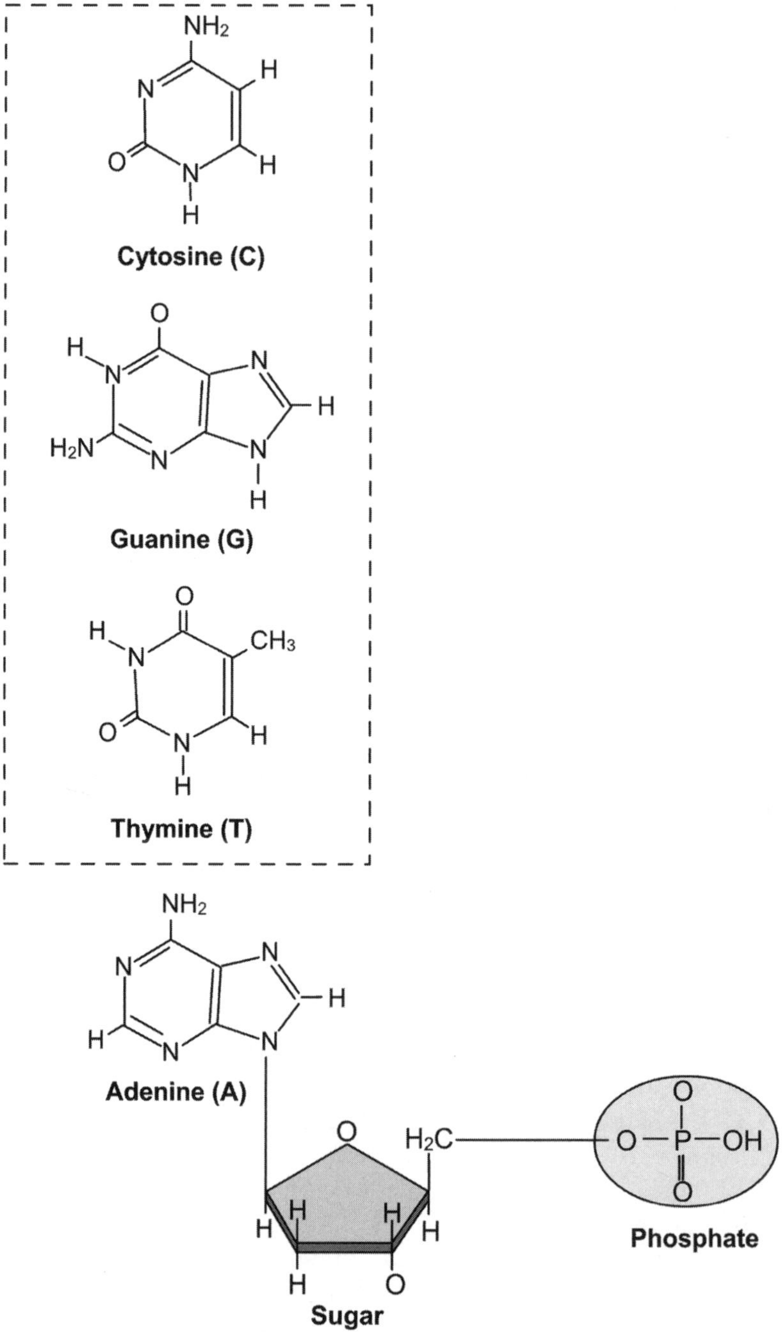

Figure 2.1 Structure of a Nucleotide. The bases within the dotted lines represent bases that may substitute for the adenine in the diagram (*Courtesy of Ricochet Productions*).

can be used as diagnostic tools to determine the presence of a specific compound in a sample.

After examining the data, Chargaff discovered some interesting patterns in the percent of each nucleotide in the tissues. First, the percent of each base did not vary significantly from different tissues of the same species, indicating that the DNA of each tissue was chemically the same. Second, for any species, the relative percent of adenine always closely matched the percent of thymine. The same was true for quanine and cytosine. It was possible for the percents of adenine-thymine, and cytosine-guanine, to vary between species, but not within members of the same species.

What Chargaff had discovered was the *complementary* nature of nucleotides. Somehow within the DNA a system was established so that the levels of adenine and thymine, and cytosine and guanine, were kept at the same levels. Chargaff did not have an explanation for this observation, but his work would be crucial for establishing an important component of DNA structure.

Wilkins, Franklin, and DNA Crystals

A key piece of the puzzle came from a group of researchers headed by Maurice Wilkins (1916–2004) at King's College in London, England. One of the members of that lab was an exceptionally bright researcher by the name of Rosalind Franklin (1920–1958). Wilkins and Franklin were interested in a field of science called biophysics. Biophysicists study the physical structure of biologically important molecules, such as proteins or DNA. One of the techniques used by Wilkins and Franklin to determine the structural characteristics of a DNA molecule was X-ray diffraction. Invented in the 1940s, the principles behind X-ray diffraction are very similar to those used in X-ray procedures by physicians. In X-ray diffraction (see Figure 2.2), a narrow beam of X-rays is focused on a crystallized sample of a molecule. As the X-rays pass through the sample, they are deflected (or diffracted) by the atoms in the crystal. The amount, and angle, of diffraction can then be recorded on a photographic plate. Crystals are preferred in this procedure over single molecules, since the repetitive nature of the crystal makes it more likely that the X-rays will be diffracted and form a clearer image on the photographic plate.

What makes this process difficult, especially with the equipment available in the 1950s, is the preparation of the molecule for study. Rosalind Franklin was well skilled in the techniques necessary to prepare molecules for study. While other researchers were using the same technique to examine DNA, it was Franklin's abilities as a scientist

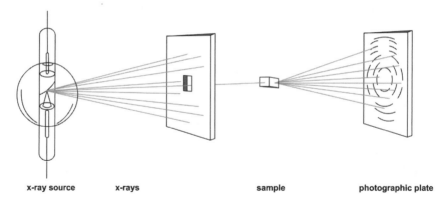

x-ray source   x-rays   sample   photographic plate

Figure 2.2 X-ray Crystallography (*Courtesy of Ricochet Productions*).

that led to her success. Through experimentation, she recognized that the best images were obtained using high-humidity environmental conditions, or "wet" DNA. She then optimized the conditions to produce some of the most detailed images of DNA structure to date.

An analysis of the images revealed several important facts on the structure of the DNA molecule. First, the images suggested a helical shape, much like that of a spring. Second, it was obvious that the molecule was highly repetitive. Third, by examining the angles of diffraction of the patterns on the photographic plates, it was possible to measure the physical characteristics of the molecule, specifically the distance between the turns of the helix and relative positioning of the atoms. There is a considerable amount of controversy on how Franklin's images were made available to James Watson (1928–) and Francis Crick (1916–2004), the codiscoverers of DNA structure (see "Watson and Crick Unveil the Double Helix" in this chapter). Most accounts report that Wilkins shared Franklin's work with Watson and Crick without her permission. Regardless, it is evident that the X-ray diffraction work coming out of the Wilkins–Franklin lab played a crucial role in the ultimate discovery of the structure of DNA. However, there were other teams racing to discover the structure of DNA, and even if Franklin's data had not been shared, the scientific community was rapidly accumulating theories and models on how the genetic material was assembled.

Pauling and Proteins

The leading investigator of a second team of researchers who were interested in determining DNA structure was Linus Pauling (1901–1994). Pauling is widely recognized as one of the leading scientists in the study of structural chemistry, specifically in the field of how chemical bonds influence the shape of a molecule. Like many of the structural chemists of the time, he used X-ray diffraction studies of crystals to determine the structure of biologically important molecules.

One of Pauling's greatest contributions toward the determination of DNA structure stemmed from his work on helical patterns in fibrous proteins. Pauling's protein of choice was keratin, a protein that forms the structure of feathers and hair. Pauling's work demonstrated that **hydrogen bonds** between atoms could influence the shape of the molecule, producing a regular pattern of bending. This pattern is now called the alpha-helix. Pauling not only identified the structure, but performed some of the most accurate calculations to date on the distances between atoms in the helix. In addition, from his study of antibody structure, Pauling suggested that it was possible that the structure of large molecules, such as antibodies, could be the result of complementary characteristics at the molecular level. However, while Pauling recognized the importance of complementation, it is unclear as to whether he understood its significance in the structure of DNA. However, his alpha-helix model of folding, and the measurements of the distance, were very influential.

Pauling's studies led him to believe that DNA was actually a triple helix, or three intertwined strands. Pauling shared this information with Watson and Crick, who realized that the X-ray diffraction images prepared by Franklin contradicted the triple-helix model, and instead suggested that the genetic material consisted of two helical strands wound around one another. The next entry details this landmark discovery.

Watson and Crick Unveil the Double Helix

By the early 1950s the research performed by Chargaff, Pauling, Wilkins, and Franklin had provided the major pieces for solving the puzzle of DNA structure. All that was needed was for someone to understand how the pieces could be connected to form a molecule of DNA. This distinction would go to Francis Crick and James D. Watson, two researchers at Cambridge University. However, like Pauling, they initially were on the wrong track regarding the structure of DNA.

James Watson was an American scientist who had trained in zoology and ornithology before beginning graduate work in microbiology and genetics at Indiana University. He was regarded as a brilliant scientist who had obtained his doctorate at the age of 22. After a brief period of study in Denmark, Watson went to Cambridge to do postdoctoral work under the direction of Sir Lawrence Bragg (1890–1971), one of the pioneers in the development of X-ray crystallography and the youngest ever recipient of the Nobel Prize. Francis Crick was an English physicist who, following World War II, developed an interest in the biological sciences and changed fields to the study of biochemistry. At Cambridge he was performing his doctoral research under Bragg on protein X-ray crystallography, which combined the fields of biochemistry and biophysics.

Watson's first introduction to DNA structure occurred in 1951 when he met with Wilkins and reviewed some of Franklin's early DNA X-ray crystallography pictures. After discussing the project with Crick, the two set out to determine the structure of the DNA molecule. Both recognized that Pauling's lab probably had a tremendous lead in determining the structure, but that did not deter their desire to enter the race. Watson and Crick's initial model was a triple helix centered around a single magnesium ion. In this triple helix the molecule's sugar-phosphate backbone was orientated toward the center of the molecule, with the helices being held in place by the magnesium ions. Triple helixes were not a new concept for the structure of DNA, researchers in Pauling's lab were working on a triple-helix structure based on their studies of protein structure (see "Pauling and Proteins" in this chapter). However, based upon their understanding of Franklin's diagrams, namely her reported distances between repeats within each helix, Watson and Crick were initially very confident of their model. Unfortunately, when they approached Wilkins and Franklin to confirm their model, Franklin quickly pointed out that Watson and Crick had based their model on a "dry" version of DNA and that their model would not work in the "wet" (aqueous) environment of the cell.

In 1953, Pauling published a paper announcing that DNA was a triple helix. However, Pauling also had placed the sugar-phosphate backbone of the molecule on the outside of the helices, an error that Watson and Crick now recognized as contradicting the fact that DNA had the properties of an acid. After once again discussing their findings with Wilkins, who shared with them updated images from Franklin's work, Watson and Crick decided to focus on a double-helix molecule that had the sugar-phosphate backbone on the outside, with the bases facing inward. In this model, the helices were held together by hydrogen bonds between the bases. However, Watson and Crick initially had similar bases pairing in each strand. In other words, a thymine in one strand formed hydrogen bonds with an opposing thymine in the other strand. Once again, the researchers believed that they had found the structure of the DNA molecule.

Unfortunately, after consulting with other biochemists, they learned they had not used the correct form of the nucleotide bases in their determination of the structure. In their model, when the similar opposing bases in each strand hydrogen-bonded, it would distort the sugar-phosphate backbone. The structure that they had proposed resulted in a DNA molecule that did not fit the dimensions specified by Franklin's X-ray images. Franklin's images revealed a repetitious pattern in the backbone, and no sign of distortion. At this point Chargaff's discovery (see "Chargaff and Complementation" in this chapter) that the amount of adenine roughly matched the amount of thymine, and that cytosine matched guanine, became important. Early on in their work Chargaff's rule seemed unimportant, since the bases were located outside of the helix and did not interact with one another, but now it provided a crucial piece of the puzzle. Watson and Crick noticed that an adenine-thymine pair occupied exactly the same amount of space as a cytosine-guanine pair. So if the adenine in one strand was paired with a thymine in the other strand, then the DNA molecule would conform to Franklin's measurements. In addition, this solved the mystery of Chargaff's rule. The reason why the amount of

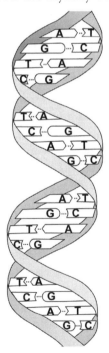

adenine and thymine were similar (as well as cytosine and guanine) was due to the base pairing between the strands.

Using Pauling's techniques of modeling protein structures, it was now possible to for Watson and Crick to present a graphic representation of their results. In 1953, Watson and Crick released a brief paper in *Nature* that described their double-helix model of DNA structure. To support their claim, Wilkins and Franklin published their X-ray crystallography work in the same issue. The structure of DNA had been unveiled. Figure 2.3 illustrates the now-familiar structure of the double helix.

In 1962 Watson, Crick, and Wilkins shared the Nobel Prize in medicine for their work in revealing the structure of DNA.

Figure 2.3 The Structure of the Double Helix (*Courtesy of Ricochet Productions*).

Unfortunately, Rosalind Franklin had died of ovarian cancer in 1958, and thus was not eligible to share in the award. While historians debate on the ethics and politics of the decision not to award Franklin a share of this important prize, many scholars agree that due to her contributions, if she had been alive Franklin would have shared the honor.

Watson and Crick's double-helix structure of DNA is worth examining in greater detail, since an understanding of the structure is crucial for comprehension of molecular genetics. First, the DNA molecule has *polarity*. This polarity is determined by the orientation of the sugar molecules in the backbone of the strand (Figure 2.4). In organic chemistry, each carbon is assigned a number. In Figure 2.4, notice how the third carbon (3′) of one sugar is attached to the fifth carbon (5′) of the neighboring sugar. DNA is said to have a *3′ to 5′ polarity* based on this configuration. The second detail regarding DNA structure that Watson and Crick determined was that the strands of a DNA molecule were orientated in opposite directions. In other words, the strands were *antiparallel*. Finally, as predicted by Chargaff's rule, the opposing strands of the double helix were *complementary*. The adenine of one strand formed hydrogen bonds with a thymine of the other strand, with cytosine doing the same with guanine. The adenine-thymine bond contains two hydrogen bonds, while the cytosine-guanine bond contains three hydrogen bonds.

MESELSON, STAHL, AND DNA REPLICATION

In the nineteenth century Gregor Mendel had demonstrated (see "Gregor Mendel and Genetics" in Chapter 1) that living organisms pass on distinct units of inheritance from one generation to another. One of the fundamental characteristics of all living organisms is the ability to reproduce. There are two basic forms of reproduction. The first is *asexual*, where the resulting offspring (or daughter cells as they are sometimes called) are genetic replicates of the parent cell. **Mitosis** is an example of asexual reproduction, as is binary fission in bacteria. The other form is *sexual* reproduction, which involves two organisms combining their genetic information for the purpose of producing an organism with a new combination of genes. For many species, this involves the process of *meiosis*, or reduction division. Regardless of the method used by a species, the parent cell must retain a copy of its genetic material. Therefore, the DNA needs to be copied, or replicated, just prior to each cell division. The replication of the genetic library must not only occur relatively quickly, since some cells have the ability to divide quickly, but also with a high degree of fidelity, since the overall purpose of cell division is to make genetic replicates, or clones, of the

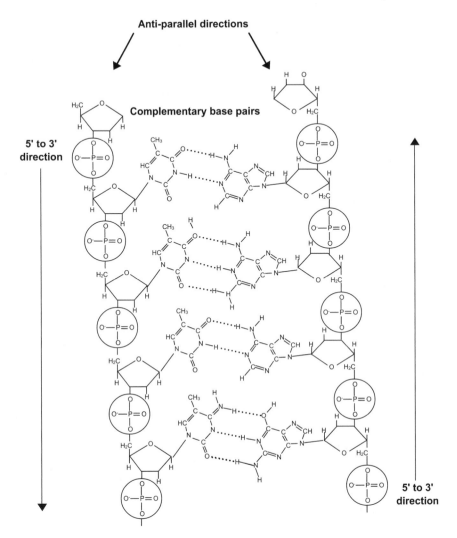

Figure 2.4 Terminology of DNA. This diagram illustrates some of the important structural terminology associated with DNA structure (*Courtesy of Ricochet Productions*).

original cell. This overall copying process is called *DNA replication,* and following the discovery of DNA structure in 1953 by Watson and Crick (see "Watson and Crick Unveil the Double Helix" in this chapter), it was possible to study the basis of replication in greater detail.

The genetic information for the formation of proteins is contained within the sequence of bases. Recall that the bases are located on the interior of the DNA molecule (see Figures 2.3 and 2.4), and thus there must be a method of accessing the interior of the strands by the replication machinery in the cell. In order to understand how DNA replicates, it was first necessary to understand what was happening to the two strands of the double helix during the copying process. There were three possibilities. First, in the *conservative* model, the original strands remained together following replication, while the newly formed strands combine to form a completely new double helix. One of the new helices is 100 percent original DNA, while the other is 100 percent newly synthesized DNA. In the *semiconservative* model, the original DNA strands separate and act as a template for the formation of a new strand. The original strand then bonds with the newly formed strand to form a double helix. Each double helix consists of 50 percent original DNA, and 50 percent newly synthesized DNA. In the *dispersive* model, the DNA strand is fragmented, and each piece acts as a template for the formation of a new strand. In this final model there is a mixture of original and new DNA in each strand. Since DNA replication produces new strands that are identical to the original, the problem was how to identify the original and old strands within each double helix.

In 1958 Matthew Meselson (1930–) and Franklin Stahl (1929–), both at the California Institute of Technology, designed an experiment to test which of these three models was correct. In this experiment they used two isotopes of nitrogen, an element that is found in the bases of each nucleotide. Nitrogen normally has an atomic weight of 14 (^{14}N), but there exists a rare form of "heavy" nitrogen called nitrogen 15 (^{15}N). Using *E. coli* as a model organism, due to its fast generation times, the researchers grew bacteria for several generations in the presence of ^{15}N only. The presence of heavy nitrogen does not influence the ability of the bacteria to function, replicate its DNA, or divide, it just becomes incorporated into the nitrogenous bases of the nucleotides generating a "marked" form of the DNA strands.

Once a population of ^{15}N-labeled bacteria was obtained, the researchers transferred the bacteria to a medium containing only ^{14}N. At specific intervals corresponding to bacterial generations, they removed samples and isolated the DNA. The DNA was then placed into a tube containing an extremely dense salt solution of cesium chloride. Next, the tubes were processed in a device called an *ultracentrifuge*, a laboratory instrument that has the ability to spin samples at speeds up to 100,000 revolutions per minute (rpm). Over a period of several days, as the machine spun the samples the heavier DNA molecules (those containing ^{15}N) would gradually separate from the ^{14}N-labeled DNA and accumulate toward the bottom of the tube.

If DNA replication were conservative, it was to be expected that the results of the centrifugation after one generation would yield two bands, one consisting of only heavy DNA (^{15}N) and one consisting of only light DNA (^{14}N). If replication was semiconservative, there should be an intermediate weight band in the first generation, since each double helix could contain both a light and heavy strand. If the dispersive model was correct, then it was unlikely that any distinct bands would be present, since each DNA strand would represent a mixture of light and heavy DNA. After a single generation of DNA replication, the isolated *E. coli* DNA presented only a single, intermediate band. This was consistent with the semiconservative model, where each strand contained both a heavy and light strand.

Meselson and Stahl's discovery of semiconservative replication is considered to be one of the classic experiments in genetics, and a milestone for the development of molecular genetics. The experimental design was not complicated, but still yielded data that clearly addressed the goals of the researchers. From this experiment, subsequent researchers were able to pursue the molecular mechanisms of how the DNA strand was replicated faithfully from generation to generation (see Chapter 3).

DNA VERSUS RNA

The earlier entries in this chapter outlined the basic role of DNA as the molecule of heredity. The molecular processes by which the information in the DNA is "read" and used to build a living organism will be explored in Chapter 3. However, DNA is not the only nucleic acid with a role in inheritance. RNA is also plentiful in all cells.

In many ways RNA is similar to DNA. Both molecules are constructed of nucleotides linked together by phosphodiester bonds. However, while DNA utilizes adenine (A), thymine (T), guanine (G), and cytosine (C) as its nitrogenous bases, RNA replaces thymine with uracil (U) (Figure 2.5). As determined by Watson and Crick (see "Watson and Crick Unveil the Double Helix" in this chapter), DNA is a double helix whose strands are held together by hydrogen bonding between the complementary base pairs (bp). In most cases, RNA is a single-stranded molecule, although some viruses possess double-stranded RNA (dsRNA). However, while RNA is single-stranded, by folding back on itself it can form complex secondary shapes that give it a variety of functions (see Table 2.3). In comparison to DNA, RNA is less stable in the cellular environment, making it a more "temporary" molecule of inheritance.

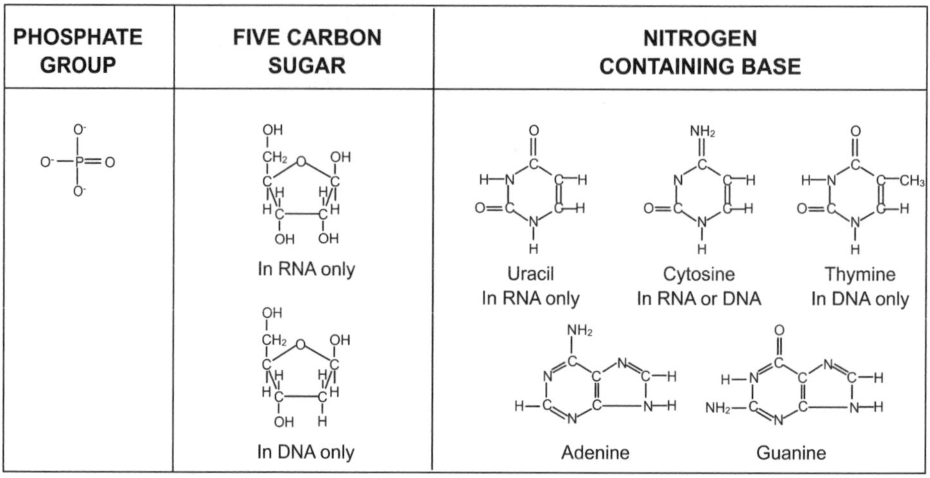

PHOSPHATE GROUP	FIVE CARBON SUGAR	NITROGEN CONTAINING BASE		

Figure 2.5 A Comparison of the Structural Units of DNA and RNA (*Courtesy of Ricochet Productions*).

The various forms of RNA play an important role in the function of the cell. The DNA can be considered to be a large, stable database of information, while the RNA may be viewed as the "working" molecule of inheritance. RNA molecules (see Table 2.3) not only play a role in the molecular processes of **transcription** and **translation**, as will be outlined in the next chapter, they also can have metabolic functions. For example, in the 1980s, Thomas R Cech (1947–) and Sydney Altman (1939–) independently discovered that RNA can function as an enzyme, a structure that is now called a **ribozyme**. These individuals were awarded the 1989 Nobel Prize in chemistry for this discovery. Ribozymes are now recognized as serving important roles for a number of cellular processes, including editing of genetic information and the synthesis of proteins. Many evolutionary biologists now believe that RNA, specifically the ribozymes, may have been the genetic material of

Table 2.3 Common Forms of RNA

Name	Abbreviation	Function
Messenger RNA	mRNA	Moves genetic information from the DNA to the ribosome, a product of transcription
Ribosomal RNA	rRNA	A structural component of the ribosome
Transfer RNA	tRNA	Involved in the process of translation
Small nuclear RNA	snRNA	RNA processing

choice in the early evolution of cells. This will be explored in greater detail in Chapter 4. Ribozymes are also being studied for use in the fields of medicine, agriculture, and biotechnology. This will be explored in more detail in Chapter 4.

3

GENETICS AT THE MOLECULAR LEVEL

Classical genetics, also called transmission genetics, examines patterns of inheritance between individuals. Gregor Mendel and the members of the Morgan lab (see Chapter 1) are excellent examples of transmission geneticists. Population geneticists examine patterns of inheritance and the movement of genes on the broader, population, scale. The study of nucleic acids, DNA and RNA, at the subcellular level is called molecular genetics. In today's scientific world molecular geneticists may be involved in a variety of fields, from ecology to medicine. However, they are linked by the fact that they are interested in how DNA contains the information for specific traits, how genes are regulated in their activity, and patterns of gene expression, to name a few. In this chapter we will explore some of the basic principles of molecular genetics. Chapter 4 will introduce the tools that molecular geneticists use to study genes in the laboratory.

MOLECULAR MECHANISM OF DNA REPLICATION

The discovery by Meselson and Stahl in 1958 that DNA replication was semiconservative (see "Meselson, Stahl and DNA Replication" in Chapter 2) opened the door for the study of how DNA is copied at the molecular level. In all organisms, DNA replication involves two primary stages. First, the strands of nucleotides within each double helix must be separated. Watson and Crick's model (see "Watson and Crick Unveil the Double Helix" in Chapter 2) had demonstrated that the bases of the nucleotide, which contain the hereditary information, were located on the interior of the double helix. The two opposing strands were held together by hydrogen bonds between the opposing base pairs, according to Chargaff's rule (see "Chargaff and Complementation" in Chapter 2). In their 1953 landmark publication in the scientific journal

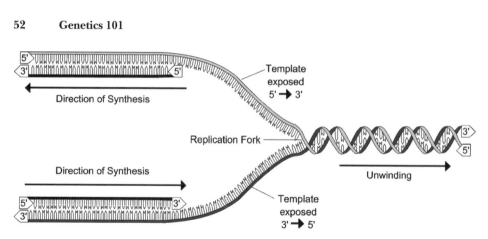

Figure 3.1 DNA during Replication (*Courtesy of Ricochet Productions*).

Nature, Watson and Crick acknowledged that the discovery that DNA is a double helix "suggests a possible copying mechanism for the genetic material." But before the information in the base pairs can be accessed, the hydrogen-bond linkages must be broken.

Base pairing does not always use the same number of hydrogen bonds. A pairing of an adenine with the thymine requires just two hydrogen bonds, while a cytosine-guanine base pair uses three hydrogen bonds (see Figure 2.3). At the molecular level this means that the adenine-thymine linkage is slightly easier to separate, and thus areas of the DNA that contain large amounts of A-T base pairs (called "AT rich" by geneticists) are ideal candidates for a location to start DNA replication. Geneticists call these areas the *origins of replication.* Bacteria, with their smaller, circular genomes, typically have a single origin of replication. Eukaryotic organisms have multiple origins of replication on each chromosome.

The method of unwinding the DNA is similar for both prokaryotic and eukaryotic organisms, although molecule geneticists give the proteins and enzymes involved slightly different names. In both systems, special proteins bind to the origins of replication and initiate the breaking of the hydrogen bonds by bending the DNA strands. Once this has begun, a special enzyme called a *helicase* is able to enter into the picture. The helicase is able to more rapidly break the hydrogen bonds holding the strands. It moves along the DNA in a 5′ to 3′ direction, forming a moving zone called the replication fork (see Figure 3.1). Once the replication fork has been formed, the other enzymes involved in DNA replication can get to work. Table 3.1 provides a summary of these enzymes.

Table 3.1 Enzymes Involved in DNA Replication

Enzyme	Function
DNA polymerase	The molecule that copies the genetic information from the template strand, forming a complementary new strand of DNA.
Helicase	Separates the DNA strands by breaking hydrogen bonds.
Primase	Manufactures small pieces of RNA that allow the DNA polymerase to begin replication.
Ligase	Connects pieces of DNA to form a continuous DNA strand.

The second stage of DNA replication involves using one of the separated strands to generate a new, complementary opposing strand. According to Meselson and Stahl, this is done in a semiconservative manner, meaning that each strand acts as a template for the formation of its complementary, or opposing, strand. The key molecule in the copying process is an enzyme called the *DNA polymerase.*

The DNA polymerase is only capable of adding new nucleotides to the 3′ end of a DNA strand. Thus, the DNA polymerase moves in a 5′ to 3′ direction. However, this presents a problem, since the DNA strands are antiparallel, meaning that the strands are orientated in opposing directions (see Figure 2.3 for review, and Figure 3.1). As the helicase and the replication fork move, one of the strands can be *continuously* copied. This strand is called the *leading strand.* On the leading strand the DNA polymerase moves along at about the same rate as the replication fork. But on the other strand the DNA polymerase is moving away from the replication fork. This means that at some point it will run out of material (DNA) to copy, and have to return to the replication fork and restart. This strand is called the *lagging strand,* and replication along it is said to be *discontinuous.* Replication along the lagging strand produces short pieces of DNA, called *Okazaki fragments,* after the Japanese biochemist Reiji Okazaki (1930–1975), that need to be joined together. The enzyme that links these strands together is called *DNA ligase.*

So how does the DNA polymerase "copy" a strand? Basically, the DNA polymerase "reads" the exposed nucleotide on the template strand, and then inserts its complement into the newly forming strand following Chargaff's rule. Thus, an exposed adenine (A) will result in a thymine (T) being incorporated into the new strand. The DNA polymerase then catalyzes the formation of the phosphodiester bond to the 3′ end of the growing strand. There are several different forms of DNA polymerases in both prokaryotes and eukaryotes, some of which have **proofreading** abilities, but their method of action is fundamentally the same.

There is another enzyme that plays an important role in DNA replication, the *primase* enzyme. The purpose of the primase enzyme is to provide a starting location for the DNA polymerase. DNA polymerases must initially bind to a double-stranded stretch of DNA, but the action of the helicase creates a single-stranded environment. The primase enzyme manufactures short sequences of RNA called *primers* that complement the exposed DNA sections near the replication fork. The DNA polymerase can then bind to the RNA-DNA segment and begin replication. The RNA primers are eventually removed by the DNA ligase enzyme.

This is just a simplified overview of DNA replication, there are other enzymes and proteins involved in the process. In addition, replication of the ends of the chromosomes (see "Structure of a Chromosome" in this chapter), requires special enzymes. This plays a special role in both aging and cancer (see "Cancer Genetics" in Chapter 7). An understanding of DNA replication led to the invention of the **polymerase chain reaction** (PCR), a technique that revolutionized the study of genetics and the life sciences (see "Polymerase Chain Reaction" in Chapter 4).

ONE-GENE, ONE-ENZYME

In the twenty-first century it is well known that genes contain DNA, that DNA is the genetic information, and that small differences in the DNA, or alleles, produce the variations in traits that give a wide range of phenotypes, including many diseases. However, following the rediscovery of Mendel's laws in 1900 (see "Gregor Mendel and Genetics" in Chapter 1) scientists began to struggle with the question of exactly what was being formed by the information located in genes. The obvious answer was the proteins, since proteins were known to be involved in practically every cellular function. Biochemists had been studying proteins for some time, but it was not until a series of experiments spanning almost 40 years that a connection was made between a protein and a gene.

The first individual to make the connection that Mendelian patterns of inheritance could be applied to the study of metabolic pathways was the English physician Sir Archibald Garrod (1857–1936). At the turn of the twentieth-century chemists had already established that many of the metabolic functions in the body were the result of chemical pathways. These started with a substrate and then went through a series of intermediate steps resulting in the formation of a final product. The chemical reaction occurring in each of the steps was performed by an enzyme. The metabolic pathway that Garrod focused on involved the dietary amino acid phenylalanine. Under normal conditions, phenylalanine is metabolized in a series of four stages, each governed by an amino acid. The metabolic pathway was well known, since a defect in one of the enzymes in the pathway would cause an accumulation of an intermediate, in much the same way as a missing worker on an assembly line halts production at that point. For phenylalanine, defects in the pathway caused a series of diseases, including alkaptonuria or black urine disease. Alkaptonuria is due to a defect in the final stage of the pathway, causing an accumulation of homogentistic acid, which turns black upon contact with air.

Through his studies of alkaptonuria in a number of families, Garrod recognized that the disease displayed a recessive pattern of inheritance. The units of inheritance are genes, and thus genes must code for the enzymes of a metabolic pathway. Garrod tested this idea out with two other diseases, cystinuria (a disease associated with amino acid metabolism) and albinism. All three diseases followed Mendelian patterns, thus confirming his ideas. Garrod called these defects *inborn errors of metabolism.* These "errors" represented mutations that limited the function of the metabolic pathway. Unfortunately, Garrod was far ahead of his time, and his ideas attracted very little attention for almost three decades.

The next major discovery that genes encode for enzymes was conducted by the team of George Beadle (1903–1989) and Edward Tatum (1909–1975). Beadle was a *Drosophila* geneticist, and following the work of Thomas Hunt Morgan and his colleagues at the fly lab (see "Thomas Hunt Morgan and the Fly Lab" in Chapter 1), many (including Beadle) believed that studies of metabolic pathways in *Drosophila*, specifically those associated with eye color, could provide an insight into the relationship between biochemical pathways and patterns of inheritance. Unfortunately, *Drosophila* did not prove to be an ideal organism for the study, so Beadle turned to Edward Tatum and his experience with the bread mold *Neurospora crassa*.

For most people the study of genetics using a bread mold might not seem interesting, but *Neurospora* proved to be the ideal organism for the study of biochemical pathways. *Neurospora* requires only a minimal array of nutrients in its environment in order to survive. *Neurospora* is able to grow on a food source containing only salt, sugar, and biotin (a **vitamin**). This is called a *minimal medium.* This means that *Neurospora* must possess the metabolic pathways to manufacture the wide array of organic compounds it needs from this limited supply of starting materials. Tatum and Bell focused on the metabolic pathways associated with growth, such as those providing for amino acid synthesis, or vitamin formation, since an adequate supply of these compounds is essential for growth, and any defect in one of these pathways would severely limit the ability of *Neurospora* to grow.

The experimental design for Tatum and Bell's experiment was relatively simple. They exposed *Neurospora* to radiation (in this case, X-rays) to induce mutations in its genetic information. If genes encoded for enzymes, then it was possible that a mutation might render one of the metabolic pathways inoperable. In this case, *Neurospora* would not be able to grow on minimal medium. However, if the medium was supplemented with an amino acid or vitamin that could not be synthesized due to the defective pathway, then *Neurospora* should be able to continue normal growth. It should be noted that Tatum and Bell did not have the means of causing mutations to occur in specific metabolic pathways, and thus their mutations occurred completely at random in the genome. Thus, they had to screen large numbers of strains (over 2,000) to find three strains that could not grow on minimal medium, but could grow on medium supplemented with a specific amino acid or vitamin (for example, vitamin B_6 or thiamine). Tatum and Bell had demonstrated that genes encode for enzymes. Tatum and Bell published their studies in 1941 and their work quickly became known as the *one-gene, one-enzyme hypothesis*. For their work Tatum and Bell received the 1958 Nobel Prize in physiology.

In the strictest sense, Tatum and Bell were not entirely correct. If they had been correct, then the over 200,000 proteins in the human body would require over 200,000 genes. We now know that the human genome contains fewer than 25,000 genes. Thus, there must be a way for a gene to code for more than one functional protein. However, the work of Tatum and Bell was important in that it established that genes code for the information in proteins. Most geneticists now state that a gene contains the information for a **polypeptide**, which may then be modified to form different versions of a protein. This will be explored in more detail in the entry "RNA Editing" in this chapter. Furthermore, not all proteins are enzymes. Some proteins form structural molecules, such as keratin in the skin. For these reasons Tatum and Bell's work is now frequently referred to as the *one-gene, one-polypeptide hypothesis*.

STRUCTURE OF A GENE

The term "gene" was first introduced in 1909 by Wilhelm Johannsen (1857–1927). At the time, the term gene was simply being used to describe the object of study of the new field of genetics. The idea that genes might be made of DNA was not evident until after the work of Avery and his colleagues (see "Avery and the Transforming Principle" in Chapter 1), and the discoveries that DNA directs heredity in viruses (see "DNA or Protein? The Hershey–Chase Experiments" in Chapter 1). Following these experiments, Beadle and Tatum demonstrated that each gene codes for an enzyme (see "One-Gene, One-Enzyme" in this chapter). Finally, Watson and Crick allowed us to visualize how DNA is the genetic material (see "Watson and Crick Unveil the Double Helix" in Chapter 2). But even after decades of research, geneticists still argue over the finer points of what makes a gene. Perhaps the easiest way to understand what is meant by the term *gene* is to place a gene within the structure of the DNA.

DNA consists of two strands of interwoven nucleotides linked by phosphodiester bonds (see "Watson and Crick Unveil the Double Helix" in Chapter 2). The sequence of nucleotides contains the "code" for making proteins, a concept that will be explored more fully in the next three entries ("Transcription," "The Genetic Code," and "Translation" in this chapter). The two DNA strands are complementary, an adenine on one strand is matched with a thymine on the opposing strand. The same is true for guanine and cytosine (see Figure 2.3). Genes may be found on either of the two strands.

Most geneticists will define a gene as the DNA sequences that are required to produce a polypeptide chain. This includes both the regions of DNA that contain the actual instructions, called the *coding regions,* for the sequence of amino acids in a protein, and the stretches of DNA sequence that help regulate the expression of the gene. These are usually called the *noncoding regions* and they play an important role in whether the gene is "on" or "off." This is called gene expression. Thus, the noncoding regions serve to determine when the information in the coding regions is expressed. Not all genes are expressed throughout a cell's, or individual's life, since not all proteins are required by every cell type.

Most texts refer to DNA as the "book of life." If we use the analogy between printed material and DNA, then a gene would be all of the information necessary to make a functional sentence. Unfortunately, this can be misleading, since we have been taught from a young age on how to "read" our language. Notice that to produce a sentence we require special punctuation, capitalization, and sentence structure. Furthermore, we can use almost endless combinations of the 26 letters in our alphabet in order to form the words of the sentence. However, if we were to look at a stretch of DNA we would observe a string of linked nucleotides containing identical sugar subunits and varying only in the type of base they possess. The "alphabet" of DNA contains only four letters, each corresponding to one of the bases—A for adenine, C for cytosine, G for guanine, and T for thymine. From these four bases we can construct something as complex as a living cell or even an organism containing trillions of cells! In addition, there is no obvious punctuation within the DNA, meaning that there is initially no apparent place to start reading the information. To complicate matters even further, genes are not lined up end to end. There can be long stretches of DNA that do not code for proteins, followed by dense clusters of genes. Obviously, there must be some method of identifying the start of a gene in this vast expanse of nucleotides. That is the role of the regulatory regions.

Regulatory regions serve to mark the start locations for genetic information that is needed to form a protein, and determine whether a gene is on or off. In most cases, regulatory regions are located before the start of the coding regions. Geneticists use the term *upstream* to indicate this directional information. The most important of these regulatory regions is the **promoter**. The promoter is located upstream of the coding regions. Think of the promoter as a light switch. Some promoters act as simple on/off systems, while others interact with other regulatory regions to vary gene expression in much the same way as a variable light switch. The role of the promoter in activating a gene will be covered in more detail in the entry "Transcription" later in this chapter.

To identify a promoter, or any regulatory region, geneticists look for patterns of nucleotides called **consensus sequences**. Not all promoters have exactly the same pattern of nucleotides, but they contain specific areas that tend to use the same patterns of nucleotides. Thus, if a geneticist examines a group of known promoters, it is possible to determine an "average," or consensus, sequence. Figure 3.2 shows how a consensus sequence can be derived from a group of bacterial promoters.

Once a consensus sequence has been determined, geneticists can use computer programs to scan the genome for these patterns, which in turn suggest that a gene may be close by.

In eukaryotic systems, including humans, there may be additional regulatory regions located upstream of the promoter. Two of these are the *enhancer* and the *silencer*. Enhancers and silencers typically act as fine tuning mechanisms for gene expression. As their name suggests,

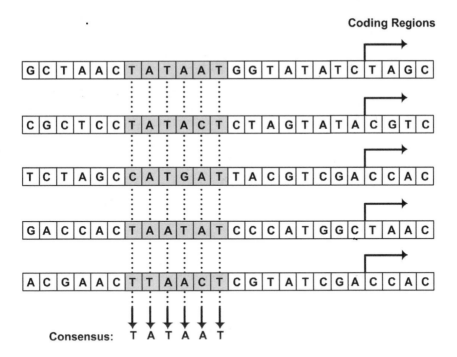

Figure 3.2 Consensus Sequence of Bacterial Promoters (*Courtesy of Ricochet Productions*).

enhancers increase gene expression, while silencers (also called *suppressors*) decrease gene expression. What is interesting is that these regulatory regions may sometimes be located at some distance from the promoter (see Figure 3.3) and exert their effect by introducing complex bending of the DNA. Scientists are just beginning to understand the complex interactions of these regulatory regions in gene expression.

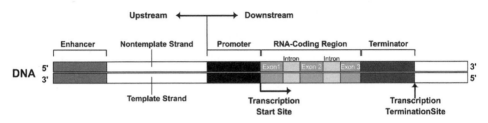

Figure 3.3 General Structure of a Eukaryotic Gene (*Courtesy of Ricochet Productions*).

The purpose of a gene is to contain the instructions for the construction of a polypeptide, a string of linked amino acids. Polypeptides are the precursors to functional proteins. This information is contained within the coding regions of the gene. The start of the coding region is located just downstream of the promoter. The next entry, "Transcription" will examine a few of the characteristics of this starting point.

The coding region of eukaryotic genes is not continuous. Instead, within a gene there are stretches of nucleotides that do not contain information. These intervening regions are commonly called **introns**. Introns interrupt stretches of coding regions called **exons**. Introns contain nucleotides, and these nucleotides are no different than the ones found in the exons. Therefore, there must be a method of removing the information found in the introns before the final assembly of a polypeptide. This editing process is not performed on the DNA, but rather on an intermediate molecule called messenger RNA (mRNA). The mRNA is altered by a process called **RNA editing**. This will be covered later in this chapter.

TRANSCRIPTION

The sequence of nucleotides in the gene represents the *genotype* of the organism, while the proteins formed by the genes control the *phenotype* of the organism. The flow of information from genotype to phenotype is a multistage process. The first step is *transcription*, during which the information in the gene is copied into an intermediate molecule of RNA. This is followed by *translation*. Translation interprets the genetic information into a string of amino acids, forming a polypeptide. Francis Crick (see "Watson and Crick Unveil the Double Helix" in Chapter 2) called this the **central dogma** of biology. The central dogma represents the primary method by which living organisms can interpret their genetic information into functional proteins. There are a few minor deviations, such as with the viruses, but overall, the system is the same for all life on Earth. The central dogma (see Figure 3.4) can also be used to illustrate why a two-step process is needed. DNA is a stable, long-term storage center for the genetic information. Think of it as an immense database. Within this database are programs, or genes, for the manufacture of proteins. Transcription accesses the database and copies only the instructions for the desired genes. This information then moves into the cytoplasm of the cell, where translation occurs. As its name implies, translation translates the nucleic acid instructions into amino acids, forming a polypeptide. Translation will be covered in detail in the next entry of this chapter.

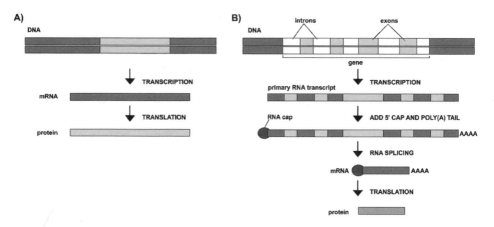

Figure 3.4 The Central Dogma of Biology (*Courtesy of Ricochet Productions*).

The process of transcription is fundamentally the same for both prokaryotic and eukaryotic organisms. The primary difference being that eukaryotic transcription occurs within the nucleus, whereas transcription in prokaryotes occurs within the cytoplasm. There are also differences in the types of proteins involved in the copying process. Overall, transcription in prokaryotes requires fewer protein assistants than the eukaryotic system. However, the general roles of the proteins are basically the same. The purpose of this chapter is not to provide a detailed examination of all of the molecules involved, but rather an overview of the process. For this reason, the assisting molecules will be described as belonging to general classes (polymerases, transcription factors, etc.).

There are three stages of transcription—initiation, elongation, and termination. The first stage, *initiation,* is the most complex of the three, since it requires not only a recognition of the correct sequences to be copied, but also a preparation of the DNA. The key event of initiation is the binding of the **RNA polymerase** to the *promoter.* RNA polymerases are complex molecules, consisting of several subunits that are assembled around the DNA in order to copy the sequence of interest. Prokaryotes typically have a single type of RNA polymerase. In eukaryotic systems, RNA polymerases exist in several forms. However, their function is all the same—to copy the DNA into a complementary RNA strand.

Promoters are "AT-rich" areas, meaning that they contain a higher percentage of adenine-thymine base-pairs than other DNA sequences.

Adenine-thymine base pairs contain fewer hydrogen bonds that cytosine-guanine pairs (see "Watson and Crick Unveil the Double Helix" in chapter 2). This means that the DNA strands in AT-rich regions are easier to separate. Separation of the strands is essential since only one strand of the DNA contains the coding sequence for the gene of interest. The promoters of eukaryotic and prokaryotic systems differ in their structure, but perform the same overall function of providing an attachment point for the RNA polymerase and its associated proteins.

The RNA polymerase does not easily bind to the promoter region, additional proteins are required to stabilize the polymerase-DNA interaction. These assistants are called either transcription factors (eukaryotes) or sigma factor (prokaryotes). Eukaryotic transcription involves a complex interaction of transcription factors, while prokaryotic systems are much simpler in design. The RNA polymerase then attaches both the DNA and the assistants. Once this has occurred, the AT-rich regions of the promoter are unwound by the polymerase. This is a similar operation to the action of the helicase in DNA replication (see "Meselson, Stahl and DNA Replication" in Chapter 2). The unwound portion of DNA is called an *open-complex*, and its formation marks the end of the initiation stage of transcription.

Once the open-complex is formed, the process of copying the DNA into a single-stranded complementary piece of RNA begins. This stage of transcription is called *elongation*. The action of the RNA polymerase in elongation is very similar to that of DNA polymerase in DNA replication, but there are a few important differences. First, the copying uses only one strand of the DNA, called the *template strand*. The other strand of DNA is called the *coding strand*, since it will have the same sequence of nucleotides as the finished mRNA. The second major difference occurs in the form of the nucleotides. In RNA, the base uracil replaces thymine. During DNA replication, an adenine (A) in one strand corresponded to a thymine (T) in the complementary strand. During transcription, an adenine in the template strand will be matched to a uracil in the newly synthesized RNA strand.

The elongation stage of transcription continues in a 5′ to 3′ direction until the RNA polymerase reaches the end of the gene. In prokaryotic genes, the end of the gene is marked by a series of nucleotides called the *terminator*. When the RNA polymerase reaches the terminator, the elongation stage ends and the *termination* stage begins. There are two methods of terminating transcription. The first involves the use of a protein, called *rho*. When the RNA polymerase reaches the end of the gene, its forward progress slows. This allows the rho protein to bind to the newly synthesized RNA molecule and dislodge the RNA polymerase from the DNA template. The second mechanism of termination utilizes a sequence of nucleotides at the end of the gene. As the RNA polymerase transcribes these nucleotides, the RNA molecule forms a loop-like structure. This structure causes an instability in the RNA polymerase, releasing it from the DNA.

In prokaryotic organisms, the end result of transcription is the production of a functional mRNA. This mRNA may then be translated into a polypeptide (Figure 3.4A). In eukaryotic organisms, transcription produces a "pre-mRNA" (Figure 3.4B). This molecule then undergoes editing (see next section on "RNA Editing") before translation.

RNA EDITING

In prokaryotic organisms and viruses there is close to a one-to-one ratio of genes and proteins. Unfortunately, the colinear relationship between the number of genes and the number of proteins does not hold true for eukaryotic organisms. For example, the human genome is estimated to contain between 23,000 and 31,000 genes, but the **proteome** is believed to contain as many as 400,000 proteins. Furthermore, studies of viral and bacterial genomes in the 1970s indicated that there was a colinear relationship between the sequence of nucleotides in the gene and the sequence of amino acids in the polypeptide. From this information it was widely believed that the output of transcription was a functional mRNA that proceeded directly to translation.

Unfortunately, when scientists began to study eukaryotic genes in the late 1970s and early 1980s, they discovered that there is not a colinear relationship between the nucleotide sequence and the amino acid sequence. Through a series of experiments, researchers were able to demonstrate that eukaryotic genes are interrupted by variable-length stretches of noncoding nucleotides called introns (see "Structure of a Gene" in this chapter). This observation, along with later determinations that the genome of eukaryotes contains far fewer genes than the proteins in the proteome, indicated that there must be a mechanism of editing the RNA prior to translation.

In eukaryotes, the end result of transcription is frequently called pre-mRNA, since most RNA molecules will undergo several stages of RNA editing prior to translation. One of these stages is **RNA splicing**, which involves the removal of the introns. The most important aspect of RNA splicing is the identification of the intron-exon boundaries in the nucleotide sequence. This is accomplished by the use of consensus sequences (see "Structure of a Gene" in this chapter) at each intron-exon boundary. The consensus sequence identifies the boundary, and the phophodiester bonds linking the nucleotides are cut at a specific location. There are several mechanisms by which the introns may be removed from the pre-mRNA. First, the RNA itself may function as an enzyme. Typically enzymes are proteins, though not always. For RNA splicing the RNA acts as an enzyme. RNA enzymes are called ribozymes, and the process by which they remove the introns is called *self-splicing.* In addition to self-splicing, pre-mRNA may be edited using an external molecule called a **spliceosome.** Spliceosomes are proteins that interact with the RNA molecule and remove the introns' sequences. After the introns are removed the exons are rejoined in order.

Another method of RNA editing is the modification of the 5′ and 3′ end of the transcript. Following transcription, a *poly-A tail* is added to the 3′ end of the transcript. A poly-A tail is a string of adenine nucleotides that can vary in size, with some RNA molecules having several hundred adenines. The poly-A tail is believed to have a role in stabilizing the RNA molecule in the cytoplasm. If the mRNA remains in the cytoplasm for a longer period of time, it may be used as a template for multiple polypeptides. The opposite (5′) end of the pre-mRNA is also modified in a process called *capping.* Capping involves the addition of a 7-methylguanosine to the 5′ end. The cap is added during transcription and appears to have a number of functions. First, the cap assists in the initiation of translation. It is also associated with the removal of introns and stabilizing the pre-mRNA molecule.

ALTERNATIVE SPLICING

Another mechanism that explains how the genome can encode for such a wide array of proteins is *alterative splicing.* Normally RNA splicing removes the introns and then reattaches the exons in sequence. However, in alternative splicing some of the exons may not be reattached. Since the exons contain coding information for the polypeptide, the removal of some of the exons will produce a slightly different form of the protein. It is important to note that the order of the exons is usually not changed, some of them are just not included in the final mRNA.

Alternative splicing is also sometimes called exon shuffling. From an evolutionary perspective, alternative splicing allows a species to produce proteins of related function from a single transcription event. Often these proteins are tissue specific, meaning that the pre-mRNA is edited one way in one tissue, but is alternatively spliced in another. There are many examples of alternative splicing in mammals, and it is believed to play a major role in explaining the size of the proteome.

———————————————————— ⟋っ◯⟍ ————————————————————

THE GENETIC CODE

Following RNA editing, the mRNA molecule contains the sequence of amino acids that are necessary for the production of a polypeptide molecule, and eventually a functional protein. However, the mRNA is simply a long chain of nucleotides, and to early researchers it was unclear as to how the protein machinery "read" this series of bases. One of the first questions was how many bases in the mRNA were required to code for an amino acid. The group of nucleotides that represent a specific amino acid is called a **codon**. Twenty different amino acids are used to make a polypeptide, therefore there must be a minimum of twenty different codons. Based upon this, it is not possible for the four nucleotides to represent an amino acid, since that would leave sixteen amino acids without codons. If a codon contained two nucleotides (AA, AU, AC, AG, etc.), that would provide codons for sixteen (4^2) amino acids. This would still leave four amino acids without codons. However, if a codon contained three nucleotides (AAA, AUA, AUU, etc.) then there would be 64 (4^3) unique codons, more than enough to cover the needed twenty amino acids.

In 1961, a research team headed by Francis Crick (see "Watson and Crick Unveil the Double Helix" in Chapter 2) became the first to experimentally determine that each codon contains three nucleotides. Following this verification, several teams set out to determine specifically what codon represented each amino acid. One of the first successes was achieved in 1961 by Marshall Nirenberg (1927–). Using an enzyme that could form RNA molecules, Nirenberg and his associates were able to synthesize an RNA containing UUU. In order to test which amino acid was coded by UUU, Nirenberg added the RNA to a series of tubes containing a protein-synthesizing system (see next section on "Translation") and one of twenty radioactively labeled amino acids. Their results indicated that protein synthesis only occurred in the tube containing radioactively labeled phenylalanine. From these results Nirenberg and his colleagues were able to determine that the codon UUU coded for the amino acid phenylalanine (see Figure 3.5). The experiment was then repeated for AAA (lysine) and CCC (proline).

Determining the amino acid for the codons was not going to be as easy. To examine the other codons Nirenberg constructed copolymer

Second Letter

		U	C	A	G	
First Letter	**U**	UUU ⎤ Phe UUC ⎦ UUA ⎤ Leu UUG ⎦	UCU ⎤ UCC UCA ⎦ Ser UCG ⎦	UAU ⎤ Tyr UAC ⎦ UAA -Ochre UAG -Amber	UGU ⎤ Cys UGC ⎦ UGA -Opal UGG -Trp	U C A G
	C	CUU ⎤ CUC CUA ⎦ Leu CUG ⎦	CCU ⎤ CCC CCA ⎦ Pro CCG ⎦	CAU ⎤ His CAC ⎦ CAA ⎤ Gln CAG ⎦	CGU ⎤ CGC CGA ⎦ Arg CGG ⎦	U C A G
	A	AUU ⎤ AUC ⎦ Ileu AUA ⎦ AUG *Met*	ACU ⎤ ACC ACA ⎦ Thr ACG ⎦	AAU ⎤ Asn AAC ⎦ AAA ⎤ Lys AAG ⎦	AGU ⎤ Ser AGC ⎦ AGA ⎤ Arg AGG ⎦	U C A G
	G	UUU ⎤ UUC UUA ⎦ Val UUG ⎦	GCU ⎤ GCC GCA ⎦ Ala GCG ⎦	GAU ⎤ Asp GAC ⎦ GAA ⎤ Glu GAG ⎦	GGU ⎤ GGC GGA ⎦ Gly GGG ⎦	U C A G

*(Right margin label: **Third Letter**)*

Figure 3.5 The Genetic Code (*Courtesy of Ricochet Productions*).

RNA molecules. A copolymer RNA contains just two bases (AAC, UUG, etc.). Using the same system of radioactively labeled amino acids, Nirenberg examined the ratio of amino acids in the polypeptide to the ratio of the nucleotides in the synthetic codon. In this manner he was able to determine the base composition (not order) of several additional codons. While Nirenberg's team was well recognized as the leader in discovering codon-amino acid relationships, they had considerable assistance from other researchers who were studying RNA at the time. One of the greatest contributions came from Robert Holley (1922–1993) at Cornell University. Holley was investigating the interaction of mRNA and transfer RNA (tRNA). Specifically, he identified the **anticodon** of tRNA that makes the process of translation possible (see next section on "Translation"). Holley predicted that the codon–anticodon interaction could be used to help with deciphering the genetic code. Nirenberg's group used this information to design an experiment based on mRNA and tRNA interactions to identify an additional set of codons. Altogether, his team isolated the amino acid counterparts of 50 codons.

Another research team, led by Gobind Khorana (1922–), also used a copolymer method. However, their experimental system was not limited to the use of single codons. Instead, Khorana, an expert in polynucleotide synthesis, and his researchers developed a method for synthesizing long chains of RNA nucleotides. In their approach, they used repeats of a single codon, for example UUC. By linking the UUC together into long chains (UUCUUCUUCUUCUUC) it was possible to study the amino acids coded for by three codons at the same time (UUC, UCU, or CUU). In this case, only the amino acids phenylalanine (UUC), serine (UCU), or leucine (CUU) would be added to the polypeptide. By using combinations of repeats, Khorana and associates were able to finish their study of the genetic code. By 1968, the genetic code had been deciphered. For their work, Nirenberg, Holley, and Khorana shared the 1968 Nobel Prize in physiology or medicine.

The genetic code is often described as being both *universal* and *degenerate*. It is considered to be universal since all living organisms that have been studied fundamentally use the same code. There are a few minor exceptions in some bacteria, the mitochondria, and yeast, but overall the code is the same. This strongly suggests a common evolutionary link between all organisms. The code is also said to be degenerate since more than one codon specifies an amino acid. There are 64 possible codons, but only 20 amino acids. In Figure 3.5, notice that both UUU and UUC code for the amino acid, phenylalanine. There are many more examples of degeneracy in the code. Often the first two bases of the codon are the same, and only the third base differs. The third base is sometimes called the *wobble base*, indicating that for some amino acids it may not play an important role.

The deciphering of the genetic code is often considered to be one of the major achievements in the study of nucleic acids. With the unlocking of its secrets, it became possible for scientists to study the effects of mutations on genes and to understand how genes function.

TRANSLATION

Translation is the term used to describe the cellular process by which the genetic information contained within the mRNA is converted into a polypeptide chain. In eukaryotic systems, the mRNA has undergone RNA editing just prior to translation. However before being translated, the mRNA must first move from the nucleus to the cytoplasm. The mRNA moves out of the nucleus through the **nuclear pores**, which act as the pathway between the nucleus and the cytoplasm of the cell. Since prokaryotic organisms lack a nucleus, the process of transcription and translation can occur simultaneously. Overall, with the exception of the movement of the mRNA out of the nucleus, the process of translation is very similar between prokaryotic and eukaryotic organisms.

The site of protein synthesis is the ribosome. Ribosomes are interesting structures in that they are a combination of RNA and protein. The form of RNA that is found in a ribosome is ribosomal RNA, or rRNA. Unlike mRNA, which carries the instructions for forming a polypeptide, rRNA is not translated. Instead, rRNA is used as a structural molecule. It will combine with proteins to form the structure of the ribosome. The rRNA and ribosomal proteins combine in two pieces, called the *large* and *small ribosomal subunits.* There are slight differences in these between prokaryotic and eukaryotic organisms, but the real difference is in the location where the assembly takes place. In prokaryotic organisms, the manufacture of rRNA and its linking with ribosomal proteins occurs in the cytoplasm. In eukaryotes, the rRNA and ribosomal proteins are assembled in an area called the **nucleolus**, and are then exported through the nuclear pores into the cytoplasm.

Another molecule that plays an important role in the process of translation is the tRNA. Like rRNA, tRNA does not code for a protein. Instead, following transcription some of the nucleotides within the tRNA transcript hydrogen bonds with one another (see "The Structure of DNA" in the chapter 2 for more information), causing the transcript to bend into a unique, cloverleaf-like structure (see Figure 3.6). At one end of the tRNA molecule is a sequence of three nucleotides called an *anticodon.* As the name suggests, the anticodon is the complement of a codon in the mRNA. Ideally, each tRNA recognizes a unique codon in the mRNA. However, due to the wobble effect (see "The Genetic Code" in this chapter), there is a slight amount of variability in codon recognition by tRNAs.

At the other end of the tRNA is an area called the *acceptor stem.* The acceptor stem holds the amino acid. Each tRNA has a specific amino acid that will bind to the acceptor stem. The type of amino acid is determined by the codon–anticodon interaction of the mRNA and tRNA. Notice that the tRNA is actually the molecule that "translates" the nucleic information into a polypeptide sequence. The codon in the mRNA carries the instructions for the amino acid sequence. The anticodon of the tRNA recognizes the mRNA codon, while the acceptor stem end of the tRNA positions the correct amino acid in position for incorporation into the polypeptide chain. Once the tRNA has released its amino acid another important molecule, aminoacyl-tRNA synthetase, attaches the amino acids to the acceptor stem of the tRNA, recycling the tRNA for later use.

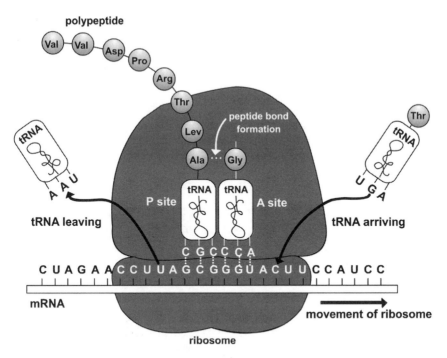

Figure 3.6 The Interaction of the Major Molecules of Translation (*Courtesy of Ricochet Productions*).

The ribosome is going to act as the organizer of translation. Its prime role is to ensure that the codons in the mRNA are being translated in the correct order. For example, in the sentence below only one sequence of letters will produce a functional sentence.

                  THEFATCATATETHEREDRAT
Correct: THE FAT CAT ATE THE RED RAT
Incorrect: HEF ATC ATA TET HER EDR
Incorrect: EFA TCA TAT ETH ERE DRA

Thus, the ribosome is going to assist in establishing the *reading frame* of the message. To do this the ribosome will interact with the tRNA and rRNA in such as way so that only a single codon may be translated at a time.

The major molecules that are involved in translation are presented in Table 3.2. The process of translation involves three stages. The first step, initiation, involves the assembly of the protein-synthesizing machinery.

Table 3.2 Molecules Involved in Translation

Molecule	Function
mRNA	Carries the genetic instructions from the DNA to the ribosome.
tRNA	Carries the amino acid to the ribosome to be incorporated into the polypeptide chain.
Aminoacyl-tRNA synthetase	Recycles the tRNA by adding new amino acids to the acceptor stem.
Ribosome	The protein synthesis structure that is a combination of both rRNA and proteins. The site of translation.
Initiation factors	Allow the ribosome to assemble around the mRNA.

Recall that the ribosome actually consists of two subunits, called the large and small ribosomal subunits. During initiation, a group of proteins called *initiation factors* assist in assembling the ribosome around the mRNA. The initiation factors temporarily recognize specific sequences in the mRNA. The small ribosomal subunit then recognizes the initiation factors, followed by the large ribosomal subunit. Thus the ribosome is assembled around the mRNA, much like a series of toy plastic blocks.

Near the beginning of the mRNA is a codon called the start codon (AUG). This codes for an amino acid called methionine (see Figure 3.5). The methionine is used to mark the beginning of the polypeptide chain and can be considered as a primer for translation. It is frequently removed from the polypeptide following the completion of translation. Three regions are important as the ribosome is assembled around the mRNA (Figure 3.6). They are commonly called the A, P, and E sites. Each site will fit a single tRNA. The only tRNA that can effectively enter the site is the one whose anticodon complements the codon of the mRNA revealed within the site (see Figure 3.6). In initiation, the assembly of the ribosome occurs with the AUG start codon within the P site. This ends the initiation stage.

The elongation stage involves the assembly of specified amino acids into a polypeptide chain. The key to elongation are the E, P, and A sites within the ribosome. Following initiation, the first tRNA (for methionine) is located within the P site. A second codon in the mRNA is exposed in the A site. Only a tRNA with an anticodon complementary to the mRNA codon exposed in the A site will correctly fit. At this point there are two tRNAs in the ribosome. By an enzymatic reaction, the amino acids between the P and A chains are joined together by a peptide bond. As the peptide bond forms, the amino acid is released from the tRNA in the P site. The ribosome then moves one codon down the mRNA (in the 3′ direction). As it does so, the tRNA that was in the P site enters into the E site and leaves the ribosome. The tRNA that was in the A site, which still has the polypeptide chain attached, moves into the P site. A new mRNA codon is then revealed in the A site. A tRNA with an anticodon complementary to the exposed mRNA codon then enters the A site, and the process repeats itself. The rate at which this reaction occurs is amazing. In eukaryotic systems, the ribosome may read up to six codons per second!

The process of termination begins once the end of the mRNA is reached by the ribosome. Notice from the genetic code (Figure 3.5) that some of the mRNA codons code for STOP signals. Stop codons are sometimes called *nonsense codons*, since they do not have a corresponding tRNA. In place of tRNAs, proteins called release factors enter into the A site. Since the release factors do not contain amino acids, the process of translation is stopped at this point. The release factors also promote the disassembly of the ribosome and its interaction with the mRNA.

An mRNA may be translated by more than one ribosome at a time. This allows the cell to make multiple polypeptide chains from a single mRNA message. The length of time that the mRNA remains in the cytoplasm for translation is determined by a number of factors, including a poly-A tail at the end of the mRNA message (see "RNA Editing" in this chapter). Thus, using RNA editing and multiple translations, a single transcription event can produce a wide range and number of proteins. The end result of translation is a polypeptide chain. This polypeptide chain must undergo a series of folds in order to produce a functional protein. This is discussed in the next section on "Protein Structure."

PROTEIN STRUCTURE

Proteins are three-dimensional molecules that perform the majority of all cellular functions. While the DNA determines the genotype, the phenotype of a cell is the result of protein action. Each amino acid contains a unique **functional group** (also called an *R group*) that establishes its chemical reactivity. As translation progresses, the functional groups in the amino acids of the polypeptide chain begin to interact with one another. As this occurs, the protein begins to bend into a three-dimensional shape. There are four levels of protein structure (Figure 3.7).

The *primary* structure of a protein represents the linear sequence of amino acids as determined by the nucleotide sequence of the mRNA. This was the information contained within the exon of the gene (see "Structure of a Gene" in this chapter). A single gene may produce polypeptides with different primary structure due to RNA editing (see

"RNA Editing" in this chapter). Few proteins retain their primary structure because as the polypeptide chain is formed during translation the functional groups of the amino acids begin to interact forming secondary and tertiary structures.

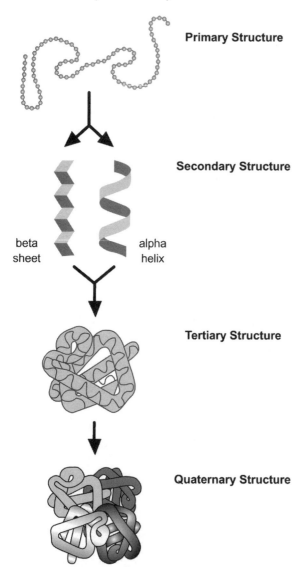

Primary Structure

Secondary Structure

beta sheet   alpha helix

Tertiary Structure

Quaternary Structure

Figure 3.7 The Four Levels of Protein Structure (*Courtesy of Ricochet Productions*).

The *secondary* level of protein structure involves a series of folds that establish a regular pattern within portions of the polypeptide chain. These patterns are the result of hydrogen bonds between the functional groups of certain amino acids. Certain patterns of amino acids result in the formation of a helical structure, called an α-helix. This is not the same as the double helix of the DNA molecule in that it does not involve nucleotide interactions or continue throughout the entire molecule. Instead, the α-helix is a region of corkscrewing in the polypeptide. Other combinations of amino acids can form a wavy sheet-like appearance in the polypeptide. This is called a β-sheet. A polypeptide can have both α-helices and β-sheets in the same protein, and multiple copies of each.

The tertiary structure is due to a variety of chemical interactions between the amino acids. The formation of secondary and tertiary structures in the polypeptide can occur simultaneously. A part of the tertiary structure is based on the interaction of the polypeptide with the water environment inside the cell. Some amino acids are **hydrophobic**, and thus prefer not to interact with water molecules. Other amino acids are **hydrophilic** and readily interact with the aqueous environment of the cell. As the polypeptide begins to bend, the hydrophobic amino acids will tend to cluster together, and the hydrophilic amino acids will orientate themselves toward the water molecules. It is important to note that this is done without breaking the peptide bonds between the amino acids. In other words, the tertiary structure can't violate the primary structure of the polypeptide. Other forces, such as ionic interactions and additional hydrogen bonding, also come into play. As the tertiary structure forms, the polypeptide assumes a complex three-dimensional shape that will ultimately determine its role in the cell.

Many proteins achieve their functionality at this level of organization. For example, pigment proteins will reflect certain wavelengths of light following tertiary folding, or enzymes gain their catalytic abilities. Minor variations in the nucleotide sequence of a gene can alter the tertiary structure slightly, producing the alleles that account for the variations in phenotype. The color of a flower may be due to a slightly different folding of a pigment protein, resulting in the reflection of different wavelengths of light and a difference in the color of the plant.

Some proteins are a combination of multiple polypeptide chains. These polypeptide interactions are called the *quaternary* level of protein structure. Not all proteins utilize quaternary structure, but those that do can sometimes form very large structures. The polypeptide chains of a protein with quaternary structure are sometimes described as *subunits*. Some quaternary structures incorporate metal ions, such as iron, manganese, and zinc, into their structure. An example is **hemoglobin**. The hemoglobin in adult blood is a large protein that contains two different polypeptides—an alpha and beta subunit. There are two copies of each subunit in a hemoglobin molecule. Integrated into the protein are atoms of iron that assist in the transportation of oxygen to the tissues of the body.

Biochemists now know that it is possible to modify proteins further by a process called *posttranslational modification*. Posttranslational modification of a protein frequently involves adding a lipid or carbohydrate group to the protein chain. This forms **lipoproteins** and **glycoproteins**, which serve a number of important functions in the cell. Other proteins may be enzymatically sliced into pieces following folding. In addition, some proteins have phosphate groups or other functional groups added. By changing the protein slightly following folding it is possible for the cell to further define the role of the protein in cellular functions.

TURNING GENES ON AND OFF

The genome of any organism contains a huge database of genes. While some of these genes are transcribed and expressed continuously throughout the life of the cell or organism, many are only for a brief period of time. In multicelled organisms, some genes are only expressed briefly during development. In fact, embryonic development in all higher organisms involves a cascade of decisions to activate or deactivate genes. In all organisms some genes are reserved for activation only during specific environmental conditions. This may allow the organism to exploit a new source of energy or render a toxin harmless. As expected, the regulation of gene expression can be very complex and research into the mechanisms of gene expressions represents a major focus of molecular biologists. New research techniques, such as the use of **microarrays** (see "Microarrays" in Chapter 4) are giving researchers the ability to study factors that activate genes under specific conditions and in certain tissues. More importantly, it allows the ability to study patterns of gene activation and interactions between genes.

An examination of a simple gene activation system in bacteria will allow us to understand the general principles of how genes are turned on and off. Our model system is the *lac* operon, which is involved in the use of the sugar lactose as an energy source in many bacteria. In bacteria genes are commonly arranged in groups called an **operon**. Genes in an operon are transcribed together but are under the control of a single regulatory region. These genes are usually functionally associated with one another. Such is the case with the *lac* operon. The *lac* operon contains three **structural genes** that allow the cell to use lactose when it is present in the environment. Some operons are normally on and continuously transcribe mRNA for protein production. They are called *repressible* operons and are turned off by a specific condition. Other operons are normally off and are turned on only when the cell needs specific proteins. This class of operons is called *inducible* operons. The *lac* operon is an example of an inducible operon.

The key to the operation of an operon are regulatory proteins, which are coded for by regulatory genes. Regulatory genes have their own promoters (see "Structure of the Gene" in this chapter), and are frequently located some distance from the operon that they are involved in regulating. The proteins produced by regulatory genes are often the mechanism by which the cell "senses" the environmental condition that is influencing the operon. In the case of the *lac* operon (see Figure 3.8),

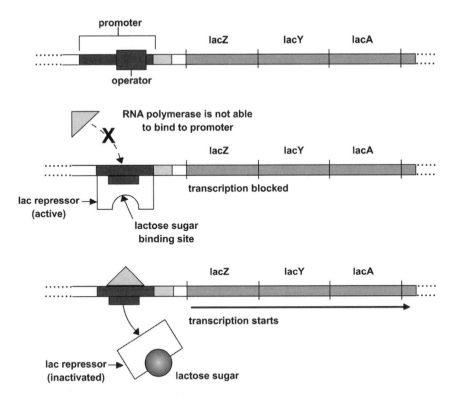

Figure 3.8 The *lac* Operon. The lac operon contains three genes, *lacZ, lacY,* and *lacA*. The regulatory gene *lacI* encodes a repressor protein that interacts with the operator of the operon (*Courtesy of Ricochet Productions*).

the regulatory gene (*lacI*) produces a repressor protein. The *lacI* gene is continuously transcribed, so the repressor protein is always present in the cell. Under normal cellular conditions, the repressor protein binds to a regulatory region in the operon called an **operator**. The operator is usually located close to the promoter of the gene. In transcription, the RNA polymerase, and its associated proteins, must bind to the promoter and unravel the DNA in order to begin making an RNA transcript (see "Transcription" in this chapter). With the repressor bound to the operator, the RNA polymerase can't initiate transcription effectively, and the operon remains functionally off.

As mentioned, the three genes in the *lac* operon allow the cell to use lactose as an energy source when lactose is present in the environment. Normally the repressor protein is bound to the operator, prohibiting the transcription of the operon. However, when lactose is present in the environment some of it is moved into the cell and converted to a chemical called allolactose. Allolactose binds to the repressor protein causing it to change its tertiary shape (see previous section on "Protein Structure"). The modified repressor protein can no longer bind to the operator of the operon. Because of this, RNA polymerase can transcribe the operon. The three genes in the operon increase the ability of lactose to enter into the cell and to be converted to a usable energy form. The allolactose does not permanently bind to the repressor, so once lactose levels drop less allolactose is present and the repressor becomes active again.

This is just one example of a regulatory mechanism. In this case the system acts in the same manner as the on/off switch of a light. However, in many cases of gene regulation the cell wants to vary the level of transcription somewhere between on and off. For example, in eukaryotic systems regulatory regions called enhancers increase the ability of the RNA polymerase to bind to the promoter (see "Structure of a Gene" in this chapter), thereby increasing the level of transcription and expression of the gene. Also, the methods of editing the mRNA and post-translational modification of proteins act as mechanisms to regulate gene expression (see "RNA editing" and "Protein Structure" in this chapter). Both prokaryotic and eukaryotic organisms have evolved a wide range of methods to regulate the expression of their genome.

4

STUDYING THE GENE

This chapter introduces some of the main techniques that are used by molecular biologists to study genes. The science of molecular genetics is a relatively new addition to the biological sciences. Molecular biology can trace its roots to the discovery of DNA structure by Watson and Crick in 1953 (see "Watson and Crick Unveil the Double Helix" in Chapter 2). In a little over 50 years the science has developed from a basic understanding of DNA structure, to the development of procedures that make it relatively simple to manipulate and study genes. In the past decade the science of molecular biology and genetic engineering has developed dramatically. Each year researchers improve on their methodology and develop new techniques to study gene structure and expression. The development of DNA microarrays, covered later in this chapter, is one example. DNA microarrays enable researchers to rapidly study how environmental factors, such as new drugs, influence gene expression. Other techniques have been developed to rapidly amplify specific sequences of DNA, allowing for advances in the study of the human genome and the genetic causes of specific diseases. The techniques that are outlined in this chapter represent some of the tools that researchers may use in the lab. The chapter is designed to provide an overview of how molecular biologists isolate, amplify, and study the gene.

RESTRICTION ENDONUCLEASES

The human genome contains almost 3.2 billion nucleotides. If the DNA in a single cell was unwound from the chromosomes, it would be over 6 feet in length. Within the genome of a species is a massive library of information, but most researchers are only interested in studying small portions of an organism's genome at a time. To obtain smaller pieces of DNA, early molecular biologists needed a mechanism of fragmenting the long DNA strands into more manageable pieces.

The answer to the problem of fragmenting the genome is a group of enzymes called *restriction endonucleases*, also known as **restriction enzymes**. Restriction enzymes are found in many bacteria, where they act as a form of "immune response" against the DNA of invading *bacteriophages*. A bacteriophage is a type of a virus that acts by injecting its DNA into a bacteria cell and hijacking the bacteria's DNA replication machinery. Once enough new viruses have been replicated, the bacteria cell bursts (or lyses) and the new viruses move on to infect additional cells. To defend against this, some bacteria have evolved a variety of restriction enzymes that fragment the incoming viral DNA, rendering it inactive. These enzymes act by recognizing a particular sequence of nucleotides (often a **palindrome)** and then cleaving the phosphodiester bonds (see "The Structure of DNA" in Chapter 2) in the DNA strands. The term *restriction endonuclease* is derived from the fact that the enzyme cleaves the internal structure (*endo*) of nucleotides (*nucleases*). Molecular biologists call the treatment of DNA with a restriction enzyme a *restriction digest*, or just a *digest*, although the process does not resemble the digestive processes of our intestinal track.

The first restriction enzyme was isolated in the early 1970s by Hamilton Smith (1931–) and Daniel Nathans (1928–1999). By convention, the names of restriction enzymes are derived from the bacterial species from which they have been isolated. The name is formed by taking the first letter of the genus name and then the first two letters of the species name. This is followed by a letter indicating the strain used in the isolation. Finally, a Roman numeral is added to the end of the name to indicate the number of the restriction enzyme. The enzyme that Smith and Nathans isolated was from the bacterial species *Haemophilus influenzae* (strain Rd) and is named *Hin*dII. Another common restriction enzyme is *Eco*RI. Its name indicates that it is the first enzyme derived from the bacteria *Escherichia coli* strain R. Table 4.1 lists some of the more common restriction enzymes used in the study of genetics, along with their sources and recognition sequences. Since the groundbreaking work of Smith and Nathans over 400 restriction enzymes have been identified, with more being added each year. For their work Smith and Nathans shared the 1978 Nobel Prize in physiology or medicine with Werner Arber (1929–), another pioneer in the discovery of restriction enzymes.

The discovery of restriction enzymes marks the beginning of modern molecular genetics. To this day these enzymes have a wide variety of applications in genetic research. For example, restriction enzymes can be

Table 4.1 Common Restriction Enzymes

Enzyme	Source	Recognition Sequence
BamH1	Bacillus amyloliquefaciens	G*GATTC
EcoR1	Escherichia coli	G*AATTC
HindIII	Haemophilus influenzae	A*AGCTT
RsaI	Rhodopseudomonas sphaeroides	GT*AC
PstI	Providencia stuartii	CTGCA*G

*Indicates the cut site of the enzyme.

used to fragment a piece of viral DNA into specific-sized pieces for use as molecular weight markers in gel electrophoresis (see next section on "Gel Electrophoresis"). In addition, since the restriction sites for a particular enzyme occur randomly throughout a stretch of DNA, variation in these sites between individuals may be used as a form of molecular marker for the study of inheritance of specific alleles (see "Molecular Markers" in this chapter).

Another useful application of restriction enzymes is to prepare DNA for molecular cloning (see "Cloning" in this chapter). When some restriction enzymes digest DNA they do so unevenly, creating single-stranded sections of DNA called "sticky ends." An example of this process using the enzyme HindIII is shown in Figure 4.1.

If two pieces of DNA are cut with the same restriction enzyme, they will have sticky ends that complement one another. These exposed single-stranded areas of DNA will naturally anneal with one another, and the broken phosphodiester bonds (see "The Structure of DNA" in Chapter 2) between the nucleotides will be repaired by DNA ligase enzymes (see "Molecular Mechanisms of DNA Replication" in Chapter 3) in the cell. The discovery of this process allowed scientists to prepare **recombinant DNA,** or DNA that is manufactured in the lab from two different sequences of nucleotides (see "Recombinant DNA" in Chapter 6).

GEL ELECTROPHORESIS

When studying genes or DNA fragments, molecular geneticists frequently use a procedure called *gel electrophoresis* to separate pieces of the DNA. The basic principle that governs the process of electrophoresis is the fact that charged molecules will move when exposed to an electrical current. When molecules are exposed to an electrical field, negatively charged particles will move toward the cathode (positive terminal), while positively charged molecules will move toward the anode (negative end). The procedure is not new, scientists during World War II discovered that

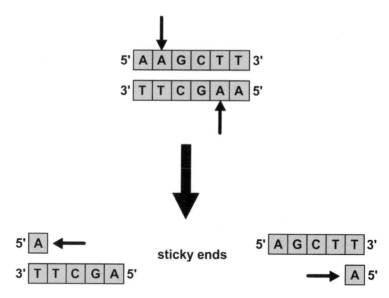

Figure 4.1 Creation of "Sticky Ends" by the Restriction Enzyme *Hind*III (*Courtesy of Ricochet Productions*).

they could separate sugars using an electrical field. The procedure has been widely used by biochemists for the separation of proteins from the 1960s onward. However, DNA is a long, linear molecule, and it wasn't until the 1970s and the discovery of restriction enzymes (see the previous section on "Restriction Endonucleases") that geneticists had the ability to generate small fragments of DNA and a need to separate and visualize small fragments of DNA.

Gel electrophoresis uses a matrix to separate the molecules. As the DNA fragments migrate through the matrix, the larger fragments will be impeded by the matrix, while the smaller fragments will pass through the pores of the matrix. The matrix that is most often used in molecular biology is agarose. Agarose is a polysaccharide derived from seaweed. The consistency of the gel determines the length of time that the gel is exposed to the electrical current. A gel with a high percentage of agarose will take much longer to run, but will provide a greater resolution of the fragments. Lower percent agarose gels will allow the fragments to move through the gel faster. Typically, gels have between 1 percent and 3 percent agarose. In addition to the percent agarose, the experimenter may vary the strength of the current. A higher current will move the fragments through the gel quicker, but will not have the resolving power of a lower current. Most genetics labs have established protocols

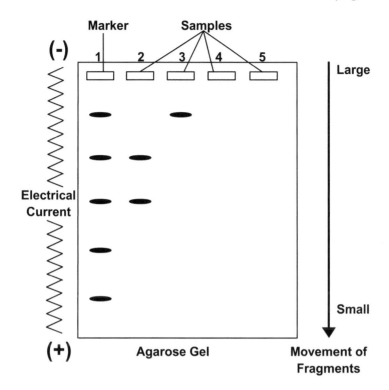

Figure 4.2 Principles of Gel Electrophoresis (*Courtesy of Ricochet Productions*).

for current and percent agarose depending on the objectives of the experiment.

An agarose gel is prepared with small indentations, called *wells,* at the end of the gel nearest the anode (see Figure 4.2). Each sample is then mixed with a loading buffer that contains glycerol and a stain, typically bromophenol blue. The glycerol helps the DNA fragments settle to the bottom of the well, while the bromophenol blue dye indicates the movement of the fragments through the agarose gel. Each sample is loaded into the gel using a micropipette, an instrument that is designed to handle samples as small as 1–10 microliters.

To determine the size of the fragments, one of the wells of the gel is loaded with a molecular weight marker. Molecular weight markers consist of DNA fragments of known size. They are usually prepared by exposing viral genomes to restriction enzymes (see the previous section on "Restriction Endonucleases"). An example is the λ—*Hin*dIII molecular weight marker. This is prepared by exposing a strain of bacteriophage called lambda (λ) to the *Hin*dIII restriction endonuclease. The result is eight fragments, with lengths of 23130, 9416, 6557, 4361, 2322, and 2027 nucleotides. Molecular weight markers, or "ladders" as they are sometimes called, come in a variety of lengths and may be purchased already prepared from a number of research supply companies.

Once the gel is loaded it is placed into a special apparatus. The apparatus is similar to a plastic food container, except that it is designed to apply a current across the gel. The gel is then submerged in a solution that contains chemicals to conduct the current and control the pH. This is commonly called the *running buffer* (or electrophoresis buffer). The electrical current is then applied to the gel through the running buffer, and the DNA fragments begin their migration toward the cathode end of the gel.

DNA fragments in an agarose gel are invisible. Therefore, the gel must be exposed to a stain in order to visualize the fragments. The most commonly used staining agent is ethidium bromide (EtBr). Ethidium bromide fluoresces a bright orange color when exposed to ultraviolet (UV) light. Ethidium bromide may be added either directly into the gel during preparation, or can be added to the running buffer during the running of the gel. Ethidium bromide works by integrating (or intercalating) between the DNA strands. It is a powerful **mutagen**, and special care must be taken in handling and disposing of this chemical. Ethidium bromide that is intercalated between the strands fluoresces slightly brighter, producing bands of light in the gel where the DNA fragments are located. Once the gel is run for a specified amount of time, it is removed from the electrophoresis apparatus and placed on a special light table that emits UV light. The results can then be photographed using black and white film. However, many labs now utilize digital photographing systems in their research. The bands in the sample DNA can then be compared to the molecular weight markers to determine the approximate size of the fragments.

Several other procedures exist to detect DNA fragments following electrophoresis, such as the use of a radioactive label. The label used in DNA studies is radioactive phosphorous (^{32}P). The radioactive phosphorous may be added to the end of a length of nucleotides. Following electrophoresis, the gel is placed with photographic film. The DNA fragments expose the film, creating bands on the film. There are also procedures that use chemical labels. However, ethidium bromide or radioactive phosphorous are the most commonly used detection procedures.

Gel electrophoresis is a useful tool for visualizing DNA fragments, but it assumes that the researcher knows some basic information about the DNA fragment, namely its approximate size. However, if genomic DNA is digested with a restriction enzyme, such as *Eco*RI, the result will be thousands of fragments of varying length. If this digest were placed in an agarose gel and exposed to ethidium bromide stain, the result would be a smear of bands. Therefore, researchers have developed methods of probing electrophoresis gels for specific fragments of interest. This process is called Southern blotting and is covered under the section "Southern Blotting" later in this chapter.

CLONING

The term *cloning* means to make an exact copy of the original. While in popular culture the use of this term is usually associated with making genetically identical sheep and humans, the process of gene cloning has been a useful tool for geneticists for decades. Gene cloning was made possible by the identification of restriction enzymes in the 1970s (see "Restriction Endonucleases" in this chapter) and the subsequent rise of recombinant DNA technology.

In theory, gene cloning is a relatively simple process. In order to make multiple copies of a gene, a researcher simply introduces a piece of DNA into a host bacterial cell. Bacteria (usually *E. coli*) are used as the host cell of choice because they are easy to culture and reproduce asexually by binary fission. Asexual reproduction does not involve the exchange of genetic information between two parental genomes, and thus the daughter bacteria cells following cell division are basically identical twins. Since bacteria divide rapidly, and must replicate their genetic material prior to each cell division, after a few generations the researcher possesses multiple copies, or clones, of the inserted DNA fragment. However, in practice the process is slightly more complicated. The problem arises with the method by which the DNA is introduced into the bacterial cells, and the means of screening large populations of bacterial cells for those cells that successfully incorporated the DNA and replicated it correctly.

The delivery of DNA into a host cell is performed using a vector. There are many different forms of vectors, but one of the preferred vectors in research with bacteria are the **plasmids**. Plasmids are small, circular pieces of DNA that operate independently of the bacterial chromosome. Plasmids are freely passed between bacteria and may exist in thousands of copies in the cytoplasm of a bacterium. Plasmids contain an origin of replication (see "Molecular Mechanism of DNA Replication" in Chapter 3), which allow them to be recognized by the host cell DNA replication machinery. Plasmids often also contain beneficial genes for the host cell. For example, many plasmids contain genes for resistance to a specific antibiotic. This property solves the second problem of DNA cloning—knowing whether the host cell contains the DNA to be cloned. If an antibiotic-susceptible strain of *E. coli* is transformed with a plasmid containing a gene conferring resistance to that antibiotic, only those bacterial cells that have successfully taken up the plasmid will be able to grow in a media containing the antibiotic.

Using these principles, the first successful cloning of a gene was done by Stanley Cohen (1922–) and Herbert Boyer (1936–) and their research associates at Stanford University in 1973. Previously, researchers had identified that certain restriction enzymes produced uneven cuts in the DNA, called "sticky ends" (see "Restriction Endonucleases" in this chapter and Figure 4.1). Using this information, Cohen and Boyer recognized that it may be possible to insert a fragment of DNA into a plasmid using a restriction enzyme. This involved several steps. First, the plasmid was exposed to a restriction enzyme, in this case *EcoRI*, which only cut the plasmid in a single location and produced sticky ends. Second, DNA of interest, which also contained *EcoRI*-generated sticky ends, was placed with the cut plasmid. The sticky ends of the DNA would complement the sticky ends of the cut plasmid. Third, by using an enzyme called DNA ligase (see "Molecular Mechanism of DNA Replication" in Chapter 3), it was possible to reform a circular plasmid, which could then be introduced back into a host bacterial cell. This altered plasmid now contains recombinant DNA, since the DNA was obtained from two different sources.

The DNA fragment that Cohen and Bayer chose to use for their first experiment was a kanamycin resistance gene. Kanamycin is an antibiotic. Thus, only those cells that had taken up the recombinant plasmid would be able to grow on a media containing kanamycin. In this experiment, kanamycin resistance served as a form of *selectable marker*, allowing the scientists to screen for only those cells that had been successfully transformed during the experiment.

Modern cloning techniques utilize many of the principles developed by Cohen and Bayer in the 1970s. The big change has been in the genetic composition of the plasmid. Modern cloning plasmids now contain two selectable markers. The first is an antibiotic resistance gene. This is typically a gene that confers resistance to the antibiotic ampicillan, although other antibiotics may also be used in the process. Only those host cells that have taken up the plasmid will be able to grow on a medium containing the antibiotic. The second marker is a gene called *lacZ*. The *lacZ* gene encodes for an enzyme called ß-galactosidase, which breaks down a normally colorless compound called X-gal into a blue product. When cells containing a *lacZ* plasmid are grown on X-gal they develop a blue color.

Table 4.2 Use of Selectable Markers in Cloning

Bacterial Colony Phenotype	Transformation[a]	Recombinant Plasmid[b]
No colonies on plate	No	?
Blue colonies	Yes	No
White colonies	Yes	Yes

[a]Transformation refers to whether the bacterial cells have taken up a plasmid containing an antibiotic resistance gene.
[b]A recombinant plasmid contains the inserted gene of interest.

Within the *lacZ* gene is a unique restriction site, meaning that it is found at only one location in the plasmid. If the plasmid is cut with this restriction enzyme, and a piece of DNA is inserted into the site, then the *lacZ* gene will no longer be functional and the host cell will lose its ability to catalyze X-gal into a blue product. Thus, the *lacZ* gene acts as a second type of selectable marker, allowing the researcher to know that they have successfully inserted their DNA and generated a recombinant plasmid. Table 4.2 summarizes the interactions of these selectable markers.

The main drawback to cloning is that the starting material for generating the recombinant plasmid is often genomic DNA. When genomic DNA is exposed to a restriction enzyme the result may be thousands of fragments, very few of which will have the gene of interest. Each of the fragments may be successfully incorporated into the plasmid and then transform a host bacterial cell. This collection of colonies is called a **DNA library**. Unfortunately, the process of cloning described above does not have the ability to discern which of the colonies contains the gene of interest. However, a library can be probed for specific sequences. This is covered in the section "DNA Libraries" later in this chapter. If the researcher wants to generate a large number of colonies containing a specific gene of interest it is first necessary to generate a large number of copies of the gene and then use this material to generate recombinant plasmids. Historically, this was only possible by conducting a series of cloning experiments where each stage screens for a select group of colonies. However, with the invention of the PCR it is now possible to generate large numbers of copies of a specific gene in a short period of time. The benefits of PCR and its interaction with the process of cloning is discussed in the next section.

POLYMERASE CHAIN REACTION (PCR)

While the process of cloning is capable of producing large copies of the gene of interest, it has several drawbacks. First and foremost is the fact that it is a time-consuming process. Growing and screening the transformed bacteria can take a considerable amount of time. Added to this is the fact that the process is not always efficient. Only a small percent of the competent cells are transformed by the plasmids, and only a subset of those will contain the sequence of interest. The next major problem is that it is not always possible to target the precise sequence of nucleotides of interest. Cloning is dependent on the location of restriction enzyme sites (see "Restriction Endonucleases" in this chapter) to define the fragments. Restriction sites are not located uniformly throughout the genome. Thus, researchers have to screen multiple clones in order to obtain the complete sequence of the gene. Cloning is useful in organisms where little is known about their genome, but by the late 1970s and early 1980s scientists were looking for a more efficient mechanism of amplifying the gene.

The structure of the DNA molecule presented the biggest challenges to developing a procedure that expedited the amplification of specific DNA sequences. DNA is a stable molecule. The DNA strands are held together by a tremendous number of hydrogen bonds (see "Watson and Crick Unveil the Double Helix" in Chapter 2). These bonds may be broken (also called denaturing) either by the use of an enzyme (such as helicase) or by extreme heat. DNA denatures at temperatures above 90 degrees centigrade. Unfortunately, at this temperature the enzymes responsible for copying the DNA denature become inactive.

However, there are organisms that can live, and thrive, at temperatures that are lethal to most other organisms. They belong in a group of bacteria called the *thermophiles* (heat-lovers), which is part of a domain of organisms called the Archaea. The thermophiles are an ancient group of organisms which can still be found in deep-sea thermal vents and volcanic fissures. Of interest to the understanding of the PCR is an organism called *Thermus aquaticus.*

In 1983, a breakthrough was made by Kary Mullis (1944–) of Cetus Corporation. Mullis realized that the secret to rapid replication of a DNA sequence involved duplicating the process of DNA replication (see "Molecular Mechanism of DNA Replication" in Chapter 3) outside of the cell and that the secret to this process was the use of a thermal-tolerant polymerase. His discovery, the PCR, centered on the ability of the polymerase from *Thermus aquaticus* to withstand the temperatures needed to denature the DNA strands. This polymerase is now called *Taq* polymerase. For his discovery, which revolutionized the study of the life sciences and genetics, Mullis received the Nobel Prize in chemistry in 1993.

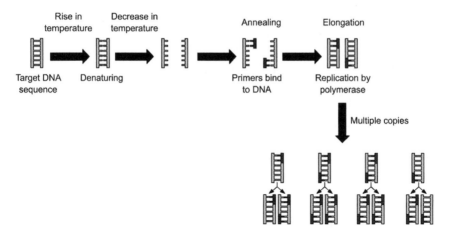

| Rise in temperature | Decrease in temperature | | Annealing | Elongation |

Figure 4.3 The Principles of a PCR (*Courtesy of Ricochet Productions*).

The principle components of a PCR reaction are a sample template DNA to be amplified, a set of primers to identify the sequence of genes to be amplified, a supply of *Taq* polymerase, and an excess of nucleotides for the synthesis of new DNA strands. The reaction is conducted in a machine called a *thermocycler*. As its name suggests, the thermocycler is designed to cycle the reaction through a series of temperatures. For the standard PCR reaction, this involves three stages (see Figure 4.3). First, the mixture is raised to a temperature in excess of 90 degrees centigrade to denature the DNA strands. Next, the mixture is allowed to cool to a temperature of around 50 degrees centigrade. As the mixture cools, the primers will bind to their complementary sequence in the DNA strands. This is called annealing. In the final stage the temperature is raised to around 70 degrees centigrade to allow the *Taq* polymerase to synthesize the new DNA strands between the two primers (see Figure 4.3). The length of time in each stage may be varied, but typically each stage takes about one minute. This completes one cycle of the reaction. In the next cycle, the newly synthesized DNA is used as template DNA. Thus, each cycle effectively doubles the amount of DNA. Most PCR reactions undergo about 30 cycles, and with modern thermocyclers can be completed in around 2 hours. From a single DNA sequence going into the reaction, the end result is over a billion copies of the target sequence. Furthermore, there is a high fidelity in the process.

There are a few variables in the process. The annealing temperature and the length of the primers determine their specificity for the target sequence. Shorter primers are less specific and will bind to more locations in the template DNA. The longer the primer, the more specific the binding becomes. In modern molecular biology labs, the sequence of the primer is generated using a computer program. In order to determine the sequence of the primers, scientists consult databases of gene sequences from similar organisms. The closer that an organism is evolutionarily, the greater the chance that they will share similar sequences. For example, prior to the completion of the Human Genome Project, researchers who designed primers for human genes could use mouse sequences as a template. The matches are not perfect, but it still allows a more rapid procedure than cloning.

DNA LIBRARIES

Even the smallest eukaryotic genome is too large to be studied as a single unit. Therefore, geneticists subdivide the genome into small fragments to facilitate the study of individual genes. Using restriction enzymes (see "Restriction Enzymes" in this chapter) and the process of cloning (see "Cloning" in this chapter), it is possible to construct a *library* of DNA fragments that can then be screened for genes of interest. The principle is relatively simple. If genomic DNA is treated with a restriction enzyme, the result will be thousands of DNA fragments of variable lengths. These can then be cloned into a bacterial cell using a plasmid vector (see "Cloning" in this chapter). Since each bacterial cell usually only takes up a single plasmid, all of the descendents of the cloned bacterial cell will have the same recombinant DNA. The bacteria are then diluted and placed on a plastic Petri dish containing a nutrient-rich medium. The dilution of the bacteria spreads the bacteria out on the plate. As the individual bacteria grow and divide on the medium it produces a small white *colony* of genetically identical cells.

To screen the colonies of a DNA library for those cells that contain the gene of interest researchers use a **probe**. Probes are short sequences of nucleotides, also called **oligonucleotides**, which are usually labeled with a radioactive isotope. The isotope of choice is usually radioactive phosphorous. As radioactive phosphorous decays, it releases small particles, called beta-particles, that are easily detected using a piece of photographic film.

To screen the bacterial colonies a piece of filter paper, usually made of nitrocellulose, is gently placed over the Petri dish. Marks are then placed on both the filter paper and the Petri dish as an orientation reference. The filter paper is then removed and placed into a container containing chemicals to break open the bacteria and denature the double helix structure of the plasmid DNA. A liquid containing the probe is then applied to the filter and allowed to hybridize with the denatured, single-stranded plasmid DNA. This is followed by a series of washes to remove any extra, nonhybridized probe. The filter paper is then placed with a piece of photographic film. The film will detect the emissions from the radioactive probe, indicating those bacterial colonies that hybridized with the probe. Samples from these colonies can then be grown in a nutrient-rich medium. When the DNA is isolated from this culture it will contain large quantities of the recombinant DNA of interest. Since the library was not destroyed in the process, it may be screened many times by repeating the above process.

There are times when a researcher is interested in creating a library of only those genes that are actively being expressed by a cell. This is especially useful in examining the response of a cell to a new drug, chemical, or environmental condition. Many genes are under the control of environmental factors (see "Turning Genes On and Off" in Chapter 3), which can readily be manipulated in a laboratory environment. One of the easiest methods of detecting the expression of a gene is to examine the mRNA produced by the process of transcription. However, it is not possible to create a library directly from RNA. Vectors, including plasmids, use DNA as the genetic material. Thus, in order to create a DNA library of these transcribed genes, it is necessary to *reverse* the process of transcription (see "Transcription" in Chapter 3). Eukaryotic cells lack the enzymes to reverse this process, but a group of viruses called the retroviruses contain an enzyme called **reverse transcriptase**. As its name implies, this enzyme is capable of converting mRNA into DNA. Since this DNA is complementary to the mRNA that is was created from, it is called *complementary DNA*, or cDNA.

Creation of a cDNA library requires the addition of a few more steps before proceeding into cloning. First, the mRNA must be separated from the rRNA and tRNA of the cell. mRNA differs from rRNA and tRNA by the fact that it contains a string of adenine nucleotides, called a poly-A tail, at the $3'$ end of the molecule (see "RNA Editing" in Chapter 3). Next, the mRNA is exposed to the reverse transcriptase enzyme, which creates the cDNA from the mRNA templates. A solution of an enzyme called RNase is then applied to remove the mRNA, leaving the DNA to use in the cloning procedure.

Another benefit of a cDNA library is that since it is derived from mature, edited mRNA, the DNA sequences will not contain any introns. Introns are only present in eukaryotic cells and thus when cloning the intron can present problems for the replication machinery of the prokaryotic cell. When using a cDNA library, the introns have already been removed by splicing, and thus only those nucleotides that are part of the coding sequence are cloned.

SOUTHERN BLOTTING

In order to gather more information about a gene, a researcher can use a process called Southern blotting. Developed by Edwin M. Southern (1938–) in 1975, Southern blotting is a procedure that uses the processes of gel electrophoresis (see "Gel Electrophoresis" in this chapter) and DNA probes (see "Cloning" in this chapter) to obtain more information about a gene, namely the size of the fragment that it resides on after genomic DNA is exposed to a specific restriction enzyme (see "Restriction Endonucleases" in this chapter). When genomic DNA is cut using a restriction enzyme, the result is fragments of varying length. When this cut DNA is separated using gel electrophoresis, and subsequently stained using ethidium bromide (see "Gel Electrophoresis" in this chapter), the result is not a series of distinct bands, but rather a smear of DNA fragments. The Southern blot transfers the fragmented DNA in the agarose gel to a nitrocellulose membrane so that it can be probed using a DNA probe that has been labeled with ^{32}P, a radioactive phosphorous isotope.

There are several important steps in the Southern blot procedure. Samples are placed in an agarose gel, which typically contains a high concentration of agarose, to provide adequate separation of the fragments (see "Gel Electrophoresis" for an explanation). Once the gel electrophoresis process is complete, the gel is exposed to a sodium hydroxide solution that denatures the double helix structure of the DNA into a single-stranded molecule. Single-stranded DNA (ssDNA) is necessary to allow binding of the nucleic acid probe later in the procedure.

The gel is then transferred to a piece of nitrocellulose paper using either pressure or a vacuum. Once the transfer is complete, the membrane is placed in a special oven to permanently bind the DNA fragments to the membrane. A DNA probe is then prepared using a short stretch of nucleotides (*oligonucleotides*). The probe may be generated from a cloned and sequenced similar gene in another species, or from a cloned gene within the same **gene family** (see later). The probe is then placed in a solution and applied to the prepared membrane. The environmental conditions under which the probe is applied to the membrane will determine how effective the probe will be in binding to its target DNA sequences. Low stringent conditions, which typically consist of high salt concentrations and low temperatures, will allow the probe to bind to DNA fragments that are not quite a perfect match for the probe. This is especially useful if the researcher is looking for anything similar to the probe, or is using a nucleotide sequence in the probe which is derived from an organism that is evolutionarily distant from the species being studied. The opposite of this is high stringent conditions. High stringency involves high temperatures and low salt concentrations. Under these conditions the probe will only bind to those fragments on the DNA that are complementary to the nucleotide sequence of the probe. A researcher can vary the conditions between high and low stringency depending on the needs of their project. The excess probe is then washed off and the nitrocellulose membrane is placed with a piece of photographic film. When developed, the film will provide an indication of those areas of the membrane where the probe bound to a DNA fragment.

Southern blots are also useful in determining whether a gene exists in multiple copies in the genome. Many genes are part of gene families, or a group of genes with similar nucleotide sequences and function. Gene families are believed to be derived from a single ancestral gene that was duplicated (see "Changes in Chromosome Structure" in Chapter 5). The duplicated genes then diverged slowly by mutation until they possessed functions but slightly different nucleotide sequences. An example of a gene family are the globins which contain the genes for hemoglobin and myoglobin, two important proteins in oxygen transport.

There are several variations of the Southern blot. Each of these have been assigned a name that is a play on the name Southern. For example, the Northern blot (originally called a "reverse-Southern" blot) separates RNA using gel electrophoresis and then probes the membrane with an RNA (sometimes DNA) probe. Northern blots are useful in studying patterns of gene expression, although they have been widely replaced by the use of microarrays (see "Microarrays" in this chapter). Biochemists frequently use a variation called the Western blot to study the proteins in a cell or tissue. Unlike the Northern and Southern blots, a Western blot uses antibodies to indicate the position of a protein on the membrane. Table 4.3 summarizes a few of the more common forms of blots and their application in the study of molecular biology and biochemistry.

DNA SEQUENCING

Ultimately, the majority of geneticists are interested in knowing the precise sequence of nucleotides in the gene that they are studying. They may be looking for specific mutations, patterns within the nucleotide sequences, or evolutionary differences between different species. However, in order to know the nucleotide sequence, researchers need a large number of copies of the sequence being studied and a method of

Table 4.3 Commonly Used Blots in Molecular Biology

Name	Target[a]	Probe	Applications
Southern	DNA	DNA (RNA)	Determine gene number and restrict fragment size.
Northern	RNA	RNA (DNA)	Gene expression by mRNA levels.
Western	Protein	Protein (antibodies)	Presence and abundance of specific proteins.
Southwestern	Protein	DNA	DNA-protein interactions.

[a]The target is the molecule that is separated by gel electrophoresis.

interrupting the nucleotide sequence at specific points so as to generate fragments of DNA with varying lengths. These fragments can then be separated using procedures such as gel electrophoresis (see "Gel Electrophoresis" in this chapter). By the early 1970s it was possible to generate sufficient quantities of a specific DNA fragment using the process of DNA cloning (see "Cloning" in this chapter). The development of the PCR in the 1980s greatly enhanced the ability of researchers to obtain identical copies of a DNA fragment quickly (see "Polymerase Chain Reaction" in this chapter). The development of new techniques in the 1970s, coupled to an increased understanding of the molecular basis of DNA replication (see "Molecular Mechanism of DNA replication" in Chapter 3), enabled researchers, such as Frederick Sanger (1918–) to develop a method of determining the precise sequence of nucleotides in a small strand of DNA.

During DNA replication, the DNA polymerase enzyme links the 5′ end of a free nucleotide to the 3′ end of the existing nucleotide strand. The bond that is formed, called a phosphodiester bond, links the hydroxyl group (-OH) of one nucleotide to the phosphate group of the new nucleotide. The chemical environment of the cell supplies the polymerase with a complete mixture of each of the four nucleotides (A, C, G, and T). The same process occurs during the PCR, except that in this case the researcher provides the mixture of free nucleotides for use by the polymerase. One of the methods of DNA sequencing acts by interrupting the ability of the polymerase to add new nucleotides at specific points along the DNA chain. This procedure, developed by Fred Sanger, is often called the **dideoxy method** of DNA sequencing (or Sanger method).

In the dideoxy method a chemically modified version of a nucleotide, called a *dideoxynucleotide* (ddNTP), replaces one of the nucleotides in the chemical mixture. A dideoxynucleotide is missing the 3′ hydroxyl group. Because of this, the DNA polymerase is unable to synthesize a phosphodiester bond between a dideoxynucleotide and the next nucleotide in the chain. This stops DNA replication at this point. For this reason this method is sometimes referred to as *chain termination.*

When sequencing DNA using the dideoxy method, a researcher prepares four identical reaction tubes. Each tube contains a single-stranded copy of the DNA to be sequenced and a primer to begin the replication reaction. In the first tube the researcher places normal A, G, C, and T nucleotides, and a small quantity of dideoxy G (ddGTP) nucleotides. The second tube contains normal A, G, C, and T nucleotides, and a small amount of dideoxy A (ddATP). This pattern is repeated for each of the remaining tubes. Once the polymerase begins replication, it will occasionally incorporate one of the ddNTPs into the strand, thus stopping replication. In the tube containing ddATP there will be an assortment of DNA fragments, varying in length, each

A
A
T
G
A
C
A
T
G
G
A
C
T
T
C

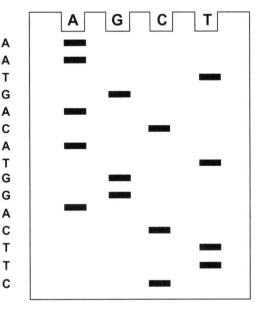

Figure 4.4 The Results of a Dideoxy Sequencing Reaction (*Courtesy of Ricochet Productions*).

with a ddATP at the end of the fragment. The four reactions are then loaded into a polyacrylamide gel and exposed to an electrical current to separate the fragments by size. This type of gel functions in the same manner as an agarose gel (see "Gel Electrophoresis" in this chapter), but runs for a longer period of time and has a finer resolution of fragment size. As is the case for agarose gels, smaller fragments will migrate further in the gel than large fragments. In order to detect the results of the reaction, the dideoxynucleotides are labeled with radioactive phosphorous (^{32}P). The gel is then placed with a piece of photographic film and developed after 24 hours of exposure. An example of the results is presented in Figure 4.4.

To read the results, start with the band at the very bottom of the gel. This represents the smallest fragment. Since it is in the C column, the reaction must have been terminated with the incorporation of a dideoxy

C nucleotide into the strand at that point. The next smallest fragment is located under the T column, indicating that the next nucleotide in the sequence is a thymine. The results are read in a ladder-like configuration up the gel. In Figure 4.4, the sequence indicated is

CTTCAGGTACAGTAA

Dideoxy sequencing reactions are limited in the number of nucleotides that can be read by a single gel. Under ideal conditions, a skilled researcher may be able to obtain several hundred base pairs with a single gel. While the dideoxy method is effective for sequencing of small stretches of DNA, such as that of a single gene, it is too time consuming for use in large-scale sequencing. A technology developed as part of the Human Genome Project was automated sequencing. Instead of labeling the nucleotides with ^{32}P, each dideoxynucleotide is labeled with a chemical that fluoresces when exposed to a specific wavelength of light. Since each ddNTP is labeled with a different fluorescent chemical, all four reactions may be loaded into the same well of the gel. At the end of the gel is a detector that records what form of ddNTP has just passed through the gel. The detector eliminates the need to "read" the results, and thus provides a faster, and more efficient, method of sequencing.

DNA MICROARRAYS

One of the newest additions in the arsenal of molecular biology techniques is the use of DNA microarrays, also known as gene chips or DNA chips. One of the main uses of a DNA microarrays is to study patterns of gene expression. Developed in the 1990s, DNA microarrays have provided a more rapid method of screening genomes for genes that are expressed under specific environmental conditions. The technology of a DNA microarray is relatively simple. Using a process called **photolithography**, which was adapted from a technology used by the electronics industry in the preparation of integrated circuits, a small piece of glass or plastic is spotted with fragments of DNA. These fragments may be genomic DNA that has been fragmented using restriction enzymes. However, most often the DNA is generated using PCR reactions so that the sequence content of each spot on the plate is precisely known. This technology allows researchers to place specific DNA sequences in specific places on the array. Thus, the sequence content of each location on the microarray is known in advance. For example, the first spot on the plate may hold the DNA sequence of an insulin gene, while the next location contains the DNA sequence of a muscle fiber gene. In the early years of development, researchers were able to place several thousand DNA sequences on a single array. However, some of the newer technologies can spot up to 1 million DNA sequences in a square centimeter area.

To use a DNA microarray a researcher must first prepare a sample of mRNA. When examining patterns of gene expression a researcher will often prepare mRNA from two different samples. For example, one may be from a normal tissue while the second sample is isolated from cells of a cancerous tissue. Once isolated, the mRNA from the tissues is then converted to a form of cDNA (see "DNA libraries" in this chapter) that contains a fluorescent tag. The cDNA is then applied to the microarray for a specific amount of time. The cDNA sequences will bind to DNA sequences on the array. After the excess cDNA is washed from the array, the chip is exposed to a specific wavelength of light that corresponds to the type of fluorescent tag used in the generation of the cDNA. Locations on the microarray that have a large number of cDNA molecules bound to them will have a greater fluorescence. This correlates to a high degree of gene expression under those environmental conditions. Often the image is digitized and the degree of fluorescence quantified for making more precise comparisons.

In the example given above the images from the normal and cancer tissue may indicate the overexpression of certain groups of genes in the cancer cell. It is useful for researchers to know whether the cancer is due to changes in a specific type of gene, such as a tumor suppressor gene (see "The Genetics of Cancer" in Chapter 7), so that treatments may be tailored for the genes involved in the cancer phenotype.

MOLECULAR MARKERS

In the early history of genetics, the inheritance pattern of a gene was traced from generation to generation based on the influence of a specific allele on a phenotype. For example, Gregor Mendel used the inheritance of pigment alleles in pea plants to derive his two fundamental laws of inheritance (see "Gregor Mendel and Genetics" in Chapter 1), while Thomas Hunt Morgan used eye color in *Drosophila* as the focus point of his studies of the chromosomal basis of inheritance (see "Thomas Hunt Morgan and the Fly Lab" in Chapter 1). However, as the study of genetics progressed, it became apparent that not all genes produced a phenotype that was easily identifiable. Many genes provide the intermediates for biochemical pathways or encode for proteins that do not directly influence a phenotype. As geneticists increasingly became more interested in the study of genomes, a need developed to establish a form of molecular "road sign" that could be used in the process of genetic mapping.

One of the biggest problems facing geneticists was how to generate these road signs in areas of the genome that did not contain genes with easily identifiable phenotypic variations. The solution is to look for patterns in the nucleotide sequences. Recall that DNA is actually a string of adenine, guanine, cytosine, and thymine nucleotides (see "The Structure of DNA" in Chapter 2). If the nucleotides are in equal proportion (25% of each), and they are distributed randomly in the genome, as is typically the case in the noncoding regions of DNA, the chances of getting a specific nucleotide at a given location is 1 in 4. Thus, we would expect an adenine to be located on average every fourth nucleotide of a DNA sequence. If we are looking for an adenine-guanine sequence (AG), then this should occur on average every 16 bases in a DNA sequence. Geneticists use the formula 4^n, where n equals the number of bases in the pattern, to determine the number of nucleotides between repeats of the pattern.

Restriction enzymes, or *endonucleases* as they are sometimes called, recognize specific patterns in the DNA strands for enzymatic cleaving. Recall from the earlier entry on restriction enzymes (see "Restriction Endonucleases" in this chapter) that each type of restriction enzyme recognizes a specific sequence of nucleotides. For example, the restriction enzyme *Rsa*I recognizes the nucleotide sequence of GTAC (see Table 4.1). Using the formula 4^n, it can be predicted that this restriction site should occur every 256 bases in a random DNA sequence. Thus, if a segment of DNA is cut using *Rsa*I, it should generate fragments that are 256 bp in length. The restriction enzyme *Eco*RI recognizes the GAATTC sequence of nucleotides. A restriction digest of DNA should produce DNA fragments of 4096 bp (4^6). Geneticists commonly round off these numbers into kilobases (kb), so *Eco*RI produces fragments of approximately 4.1 kb.

These fragments can be viewed using either gel electrophoresis or Southern blots (see sections on these procedures earlier in this chapter). However, if a sequence of DNA has a single nucleotide mutation that disrupts the recognition sequence of the restriction enzyme, then the pattern of bands on the gel will be different. For example, a mutation in an *Eco*RI restriction site would prevent the enzyme from cleaving the DNA at one location, producing a 8.2kb (4.1kb + 4.1kb) fragment. The difference between a 4.1kb and 8.2kb fragment can easily be detected using gel electrophoresis. These variations in the patterns of fragments on a gel or blot are called restriction fragment length polymorphisms, or **RFLPs**. RFLPs represent the earliest form of molecular markers to be identified.

The rapid identification of patterns in the DNA became possible with the development of the processes of the PCR and DNA sequencing (see these sections earlier in this chapter). Using PCR, researchers are able to amplify specific sections of the genome for analysis. This greatly expedited the search for molecular markers, since the researchers could focus on areas of the chromosomes that had demonstrated linkage to a disease or trait. The development of rapid and automated sequencing techniques also enhanced the ability of researchers to isolate very small molecular markers called single nucleotide polymorphisms (SNPs). SNPs, are becoming a popular form of genetic marker among geneticists. Pronounced as "snips" by geneticists, SNPs are single nucleotide variations in a DNA sequence that occur in at least 1 percent of the population. Researchers estimate that there may be over 10 million SNPs scattered throughout the human genome, making them a potentially powerful form of genetic marker.

In human genetics, SNPs are widely used in what are called **association studies**. In association studies geneticists first identify candidate genes for a disorder. Using available databases, or by sequencing the gene directly, they then locate SNPs in the population. These SNPs are usually very closely linked to the gene of interest, and may often be present in the introns of a gene (see "Structure of a Gene" in Chapter 3). DNA is then isolated from individuals who have the disorder, and a control population which does not exhibit the disorder. For each sample, the sequence of nucleotides around the SNP is amplified using the PCR and then sequenced to determine the presence or absence of the SNP. If individuals who have a specific SNP also have the disease, then the SNP and the specific allele of the gene causing the disease are linked, and the SNP can be used as a diagnostic tool for physicians and genetic counselors. This approach has been used very effectively in the study of sickle-cell anemia and other human diseases.

Genomes often have areas where specific nucleotide sequences are repeated in tandem. Frequently the number of times that the sequence repeats itself is highly variable. For this reason these patterns of repeats are called **variable number of tandem repeats**, or VNTRs. VNTRs were first described by the British geneticist Alec Jeffries (1950–) in 1984. VNTRs are easily detected by a PCR, since variations in the number of repeats will produce PCR products that vary in length. These products can be easily separated using gel electrophoresis. The analysis of VNTRs forms the basis for DNA fingerprinting, also known as DNA typing or DNA profiling (see Chapter 7).

One form of VNTR are the **microsatellites.** Microsatellites are repeated sequences consisting of one to three nucleotides. The repeats occur in tandem up to several hundred times. An example of a microsatellite in humans is the CA repeat. The CA repeat (CACACACA-CACACA) occurs frequently in our genome, on average once every 30,000 bp. Microsatellites are sometimes called simple sequence repeats (SSRs). Another form of VNTR is the **minisatellites**. The repeated unit of a minisatellite is longer than that of the microsatellite, it can be up to 100-bp long and may be repeated in tandem over a thousand times. Minisatellites are highly polymorphic in their length and are often used to identify individuals during the process of DNA fingerprinting (see "DNA Fingerprinting" in Chapter 7).

With the completion of the Human Genome Project, many SNP, RFLP, and VNTR markers have been identified. This is making it possible for researchers to quickly isolate genes of interest and determine their association with a specific trait. It has also made it possible for researchers to begin finalization of a precise genetic map of the human genome (see next section on "Genetic Maps" in this chapter).

GENETIC MAPS

The first genetic map was constructed by Alfred Sturtevant in 1911 (see "Thomas Hunt Morgan and the Fly Lab" in Chapter 2). Sturtevant was a member of Thomas Hunt Morgan's Fly Lab, a site of intense activity in the study of transmission genetics at the beginning of the twentieth century. Sturtevant, working with morphological sex-linked traits in *Drosophila melanogaster*, recognized that he could generate a genetic map of the sex chromosome by determining the percent of recombinant phenotypes in the offspring of a sex-linked cross.

The type of map generated by Sturtevant is called a genetic map in that it is constructed using recombination frequencies. Recombination frequencies are based on crossing over during prophase of meiosis. The further apart two traits are, the greater the chance that recombination will occur between the two points during meiosis. Recombination is relatively low between two closely linked traits or markers. Geneticists have established that a 1 percent recombination rate is equal to 1 map unit (mu), also called a centiMorgan (cM). Thus, if recombination is observed in 37 percent of the progeny of a cross, the traits are said to be 37 mu apart. Genetic maps based on linkage are useful in placing a gene on a general region of a chromosome, but since even 1 mu can equal 1 million base pairs (in humans), it is not highly accurate.

Overall, genetic maps are useful in establishing the position of a trait in relation to other traits or genetic markers. However, in order to be accurate a large number of progeny need to be observed in a relatively short period of time. Thus, genetic maps are often constructed in model organisms, such as *Drosophila melanogaster*, which produce large numbers of offspring in a single, short, lifespan.

A physical map is determined using sequence analysis. Physical maps determine distances between two points in terms of base pairs. Since these distances are often very great, the values are often expressed in either kilobases (kb) or megabases (mb). Thus a distance of 1,320,000 is represented as either 1320 kb or 1.32 mb. In order to construct a physical map, geneticists first divide the genome into sections by genomic cloning (see "Cloning" in this chapter) and then sequence each section individually. These segments of the genome are called *contigs*. Contigs are designed to overlap each other. Computer software then compares the overlapping regions for similar sequences and constructs a physical map of the region. The overall goal of the Human Genome Project was to construct a physical map of the human genome. The technological advances associated with this project have greatly aided the construction of physical maps in other organisms.

The third type of genetic map is the cytological map. A **karyotype** is a form of a cytological map. Karyotypes are prepared by treating chromosomes with stains to produce banding patterns. These banding patterns allow cytogeneticists to distinguish individual chromosomes and are useful in assessing the presence of chromosomal abnormalities such as insertion, deletions, translocations, or inversions (see Chapter 5). In addition to staining, cytogenetic maps may be generated using labeled pieces of DNA. For example, if a researcher has sequenced a gene of interest, and wants to know approximately where in the genome the gene is located, they may label the sequence with a fluorescent marker and then allow the sequence to hybridize with the chromosomes. This procedure is called **fluorescent in situ hybridization** (FISH). By exposing the hybridized chromosomes to a specific wavelength of light the label will fluoresce, indicating the approximate position of the gene on the chromosome.

All three of these maps can be reconciled into a single map of an organism's genome if markers can be identified that can be anchored to a specific location in the genome. These sequences are called sequence-tagged sites (STS). An STS is a nucleotide sequence that has a unique location in the genome. In some cases an STS may contain a VNTR, which allows it to be used both as a marker and an anchor. Once a position has been anchored in the genome, it may be used to correlate the physical and genetic maps.

Genetic maps exist for a variety of most of the model organisms, including *Drosophila melanogaster*, *Arabidopsis thaliana*, and the nematode *Caenorhabditis elegans*. One of the ongoing goals of the Human Genome Project is to derive a complete genetic map of the human genome. Already maps of some of the chromosomes are complete, and more detailed maps are being continuously added.

5

CHANGES TO THE GENETIC MATERIAL

In the organic world, change is inevitable, even if the molecule is as stable as DNA. Changes in the genetic information are the basis of evolution. Thus, the processes examined in this chapter represent the underlying forces behind evolutionary change. The majority of changes to the genetic information are detrimental to the success of the organism. Thus, by the process of natural selection, these individuals would be at a reproductive disadvantage, and the new variation of the gene will not be passed on to the next generation. Some changes are neutral, meaning that they do not directly influence the phenotype of the organism. However, occasionally a mutation in a gene or a chromosomal change will prove to be advantageous to an organism, either by providing it with a reproductive advantage or allowing it to adapt to a new food source or habitat. These changes, while rare, are the foundation of evolutionary change.

This chapter examines the principle mechanisms by which the genetic information may be changed at either the gene or chromosome level of organization. These changes may be the result of mutations in a gene, the rearrangement of chromosomes, or a change in total chromosome count in a cell. It also takes a look at the basic principles by which a cell prevents change using DNA repair mechanisms.

CHROMOSOME STRUCTURE

Chromosomes are the organizing structures for the genetic library of the cell. However, chromosomes are not simply a long chain of genes that are evenly spaced along the length of the chromosome. Instead, chromosomes are complex structures with distinct structural features that play important roles in the processes of replication, transcription, and regulation of gene expression. Within the chromosome are regions that are densely packed with genes, and other areas that are genetic "deserts." In addition to the genes, located throughout the chromosome are sequences of nucleotides that are repeated over and over again. The function of many of these repetitive sequences is unknown, but molecular biologists have learned to use them in a variety of ways to study the gene.

An added level of complexity for chromosomes is the fact that a chromosome is not comprised solely of DNA. DNA is a double-helix molecule. While flexible, it generally does not have the ability to form complex structures on its own. However, DNA will interact freely with special structural proteins, forming **chromatin**. Chromatin plays an important role in compacting the genetic material so that it can fit into the small space of a nucleus or prokaryotic cell. For example, in order for the single circular chromosome of a typical bacterium to fit inside the cell its size must be reduced 1000 times. This is done first by the formation of loops in the DNA, followed by a process called *supercoiling*. Loop formation and supercoiling are both made possible by special proteins that interact with the DNA and facilitate folding.

DNA compaction in eukaryotes is more complex and involves a group of proteins called the *histones*. Like the proteins involved in supercoiling in the prokaryotes, histones readily interact with DNA to promote bending. The DNA-histone complex is referred to as a *nucleosome*. Conceptually, the nucleosome resembles a yo-yo, with the DNA wrapping around a combination of histone proteins. Following this first stage of compaction, the nucleosomes interact with one another, and additional proteins, to form an increasingly dense structure. Highly condensed DNA is called *heterochromatin*. Since it is highly condensed, the DNA is relatively inaccessible. Thus, areas of the chromosome that contain heterochromatin contain few genes. In comparison, stretches of chromatin that are less compact, thus allowing for greater gene expression, are called *euchromatin*.

The majority of genes are located within the euchromatin of a chromosome. An examination of the structure of euchromatin reveals that there are *repetitive sequences* of nucleotides within the DNA strands. Within a genome, some sequences of DNA are repeated thousands or millions of times. These are called *highly repetitive sequences*, and they do not typically represent genes. Highly repetitive sequences are frequently used in processes such as DNA profiling (see Chapter 7).

Genes that exist in multiple copies within the genome comprise the *moderately repetitive sequences*. These are usually genes that are necessary in large quantities within the cell. Examples are the genes that encode for the tRNA molecules involved in translation (see "Translation" in Chapter 3), genes that code for rRNA, and the genes that provide the instructions for the synthesis of the histone proteins. In contrast, genes that code for the metabolic enzymes, or proteins that establish the phenotype of the organism, are typically only found in a few copies in the genome. These are called *nonrepetitive sequences*. Most of the genes are found in the nonrepetitive sequences of the chromosomes.

In addition to the level of DNA compaction, prokaryotic and eukaryotic organisms also have several differences in the structure of their chromosomes. Prokaryotic organisms have a single, circular chromosome, while the eukaryotes have multiple, linear chromosomes. Since prokaryotic chromosomes are circular, and contain far less genetic material than their eukaryotic counterparts, they typically contain a single origin of replication (see "Molecular Mechanism of DNA Replication" in Chapter 3) and no **telomeres**.

As mentioned, eukaryotic chromosomes are linear. They contain multiple origins of replication (see "Molecular Mechanism of DNA Replication" in Chapter 3). Replication involves the formation of two identical DNA molecules in preparation for cell division. These are called sister chromatids and they represent identical copies of the same DNA strand. Prior to cell division, or mitosis, the sister chromatids are linked at regions of heterochromain called the **centromere**. In cell division, proteins bind to the centromere forming a structure called the *kinetochore*. The kinetochore interacts with the spindle fibers to facilitate the separation of the chromatids during anaphase of mitosis. The centromere is not necessarily located in the center of the chromosome. In fact, based on the location of the centromere on the chromosome, cytogeneticsts have identified four classifications of chromosomes. These are illustrated in Figure 5.1b. The position of the centromere also divides the chromosome into two parts, which cytogeneticists call the long and short arms of the chromosome. This distinction becomes important when identifying and mapping the chromosome (see the next section on "Observing Chromosomes").

The last major structure of a eukaryotic chromosome are the telomeres. The telomeres may be considered to be the "caps" at the end of the chromosome. They consist of highly repetitive heterochromatin with very few, if any, genes. The telomeres serve to stabilize the structure of the chromosome and prevent it from degradation or translocations (see "Changes in Chromosome Structure" in this chapter). Interestingly, most cells can't replicate their telomeres, and thus after each cell division the telomeres shorten slightly. In this capacity the telomere acts

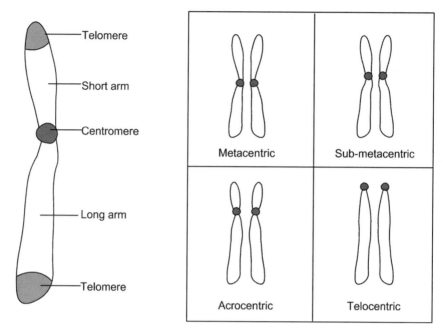

Figure 5.1 Anatomy of a Chromosome (*Courtesy of Ricochet Productions*).

as a cellular "fuse." Once the telomere shortens, the cell is no longer able to replicate, and cell death occurs. For this reason telomeres are frequently the target of aging research and the study of immortal cell lines, as is the case with most cancers.

THE ROLE OF TELOMERES IN AGING

Unlike the majority of the chromosomes, the telomeres are not completely replicated when a cell divides. In mammals, the length of the telomere shortens by about 30 to 200 bp per cell division. Since telomeres do not contain genes, and consist of repetitive sequences (TTAGGG in humans), the loss of a few base pairs per cell division does not have an immediate influence on the phenotype of the cell or its metabolic function. However, as the cell ages, and goes through multiple cell divisions, the shortening of the telomeres begins to destabilize the chromosome. The rate of shortening is not constant, meaning that researchers do not believe that it can be used as an "aging clock." This is because other factors, such as free radicals, can accelerate the rate of telomere loss. However, as the telomere shortens the cell becomes more susceptible to **apoptosis**, or programmed cell death.

Certain cells have a mechanism of maintaining telomere length. These cells contain an enzyme called **telomerase**, which serves to replicate the telomeres following DNA replication. In humans, most of the body's cells lack telomerase activity. These cells have a limited lifespan. However, in some types of actively dividing cells, such as the stem cells of the bone marrow, the telomerase enzyme is active, allowing for continuous cell division. Rapid cell division is a characteristic of cancer cells and most cancer cells have their telomerase enzyme turned on. The rapid cell division in these cells is responsible for the formation of tumors. However, the presence of an active telomerase enzyme is not a cause of cancer, other mutations must first occur in a cell for it to become cancerous.

OBSERVING CHROMOSOMES

The study of chromosomes began in the nineteenth century as part of a developing interest in the study of cell biology. Prior to this time microscopists did not possess microscopes with sufficient resolving power to accurately distinguish the interior components of the nucleus. The ability of scientists to study chromosomes was greatly enhanced by the invention of synthetic dyes in the mid-nineteenth century. Originally designed for use in the textile industry, cell biologists quickly recognized that they had tremendous applications in the study of cell biology, specifically in the study of chromosomes. This began the science of **cytogenetics,** the area of genetics devoted to the study of chromosomes.

In a diploid organism, each cell has two copies of each chromosome. However, these copies are rarely found in the same location in the nucleus. Instead, within the nucleus the chromosomes exist as a massive jumble of genetic material. During most of the life of a cell, the DNA within the chromosome is not condensed and is unraveled to facilitate gene expression. However, during cell division (either mitosis or meiosis), the chromosomes condense and become visible when stained and viewed under a light microscope. However, many chromosomes have the same basic shape and size (see the previous section on "Chromosome Structure"). In order to distinguish between the chromosomes it is necessary to use some form of a tag that produces recognizable patterns on the chromosome. Cytogeneticists frequently use chemical dyes to stain the chromosomes. Depending on the type of stain being used, the stain interacts with stretches of nucleotides that either have a high percentage of adenine and thymines (called "AT rich), a high percentage of guanine and cytosine ("GC rich"), or heterochromatin (see the previous section on "Chromosome Structure"). In most procedures the chromosomes must first be treated with heat or an enzyme to prepare them for staining. One of the more commonly used chemical dyes is Giemsa stain. Giemsa stain is a combination of methylene blue and eosin and has been successfully used for the identification of chromosome bands since the early 1970s.

Following completion of the staining process, each chromosome has a distinct pattern of light and dark colored bands. Figure 5.2 illustrates a human chromosome that has been stained with Giemsa stain. The individual bands are not genes, but rather represent regions of the chromosome. If the procedure were reproduced using a different form of stain, the result would be a different pattern of bands. In fact, cytogeneticists have a wide array of chromosomal stains. Research papers on chromosomes frequently cite the use of R or Q banding. Once stained, the chromosomes can be photographed and sorted by bands and size. This is called a **karyotype**, or *ideogram.* Historically, karyotypes were constructed by hand, with scientists cutting out individual photos of chromosomes and arranging them. However, in modern research the process has been increasingly computerized, making karyotypes easier, and faster, to obtain.

The banding patterns for each type of stain may be used to generate a physical "map" of the chromosome. The centromere (see the previous section on "Chromosome Structure") of the cell is used as a starting point. Each arm of the chromosome is then divided into major regions using the larger bands. The first major region of a

Figure 5.2 G-bands and Human Chromosome 11 (*Courtesy of Ricochet Productions*).

chromosomal arm from the centromere is labeled 1, the second 2, and so on. Within each region there are usually a number of smaller bands. For each region, these are numbered outward from the centromere. While it may appear confusing as to what defines a region and a band, scientists have established committees that review banding patterns to provide a consensus. For some genomes, such as the human genome, the banding patterns are well identified and recognized by the scientific community.

The use of chromosomal staining and karyotype analysis provides a chromosomal address of sorts. Figure 5.2 provides an example of chromosome 11. A gene located at chromosomal location 11q23 is found on chromosome 11, on the long arm (q), region 2, band 3. Using high-power microscopes, additional subbands may sometimes be detected. If so then they are identified after the band number. In our example, 11q23.1 would indicate the first subband in band 3 of the second region.

Chemical stains are not the only method of identifying and observing chromosomes. Another process uses molecular probes to study chromosome structure. A molecular probe is a short sequence of nucleotides (DNA or RNA) that will complement a specific region of a chromosome. This region can be a gene or a repetitive sequence. One sequence that is commonly tagged is the *Alu* sequence, which can occur several million times in the genome (see "Transposons" later in this chapter). The probe is labeled with a fluorescent compound. Fluorescent compounds absorb light at one wavelength, and emits light at another wavelength. In other words, a fluorescent compound "glows" when exposed to a specific wavelength of light. The use of fluorescent tags to examine chromosomes is called fluorescent in situ hybridization. In situ is a Latin term meaning "in place." The process is commonly called FISH. Once a fluorescent tag is hybridized to a chromosome, it can be observed using a special fluorescent microscope. These microscopes emit light in specific wavelengths and are frequently attached with a camera to record the results. The result is a series of fluorescent bands that researchers can utilize for chromosome identification.

FISH may also be used to determine the location of a specific gene on a chromosome. Once a gene has been isolated and amplified (see the material presented in Chapter 4) it is possible to generate a fluorescent-labeled copy of the gene and then study where it binds to in the genome. The process is not always precise since many genes contain similar sequences, but it has proven useful in determining the physical location of a gene on a chromosome.

CHANGES IN CHROMOSOME STRUCTURE

Chromosomes represent the filing cabinets for the long-term storage of the genetic information. Many organisms, including most plants and animals, are diploid, meaning that they have two complete sets of chromosomes, one obtained from the maternal line, and one obtained from the paternal line. During the formation of gametes (in animals), these chromosomes must line up with one another and exchange genetic information. The chromosomes line up with another by aligning regions of similar genetic sequences. Since each of these chromosomes is going to separate and end up in an egg or sperm cell, it is important that the chromosome align correctly so that there is an equal exchange of genetic material and a correct set of genes on each chromosome.

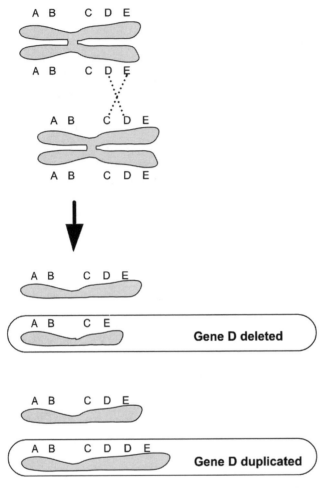

Figure 5.3 Misalignment of Chromosomes Causes Duplications and Deletions (*Courtesy of Ricochet Productions*).

However, sometimes the alignment of the chromosomes does not occur correctly. This usually occurs when there are sequences on the chromosome that are similar to one another. These are called homologous sequences, and they may either be contained within similar genes, or in the stretches of DNA between the genes. In eukaryotic chromosomes these repetitive sequences can range in size from several hundred to several thousand base pairs in length. The repetitive sequences can cause the chromosomes to misalign, as is illustrated in Figure 5.3. This misalignment is the leading cause of duplications and deletions on a chromosome.

As its name implies, a duplication causes there to be an extra copy of some of the genetic information on one of the chromosomes (see Figure 5.3). As you can observe in the figure, a duplication event on one chromosome usually correlates to a deficiency being generated on the other chromosome. Although it may not seem that having too many copies of a gene could be detrimental, at times it is. That is because genes encode for proteins (see the entries "Transcription" and "Translation" in Chapter 3), and proteins determine the phenotype of the organism.

An excess of genes may result in an overabundance of a certain protein in the cell, which in turn may alter the phenotype in an undesirable manner. However, duplications are not usually as detrimental as deficiencies.

A nice example of how additional copies of a gene can produce a detrimental phenotype is the study of the bar eye phenotype in *Drosophila*. Bar eye is a phenotype that is characterized by a reduced number of facets in the eye of the fly. Calvin Bridges (1889–1938), a member of the Morgan "fly lab" in the 1930s (see "Thomas Hunt Morgan and the Fly Lab" in Chapter 1), was the first to discover that the number of facets was related to the number of copies of the *bar* gene. Using staining techniques (see previous section on "Observing Chromosomes"), Bridges observed that an increase in the number of copies of the *bar* gene on the X chromosome resulted in a decreased number of eye facets. An overproduction of gene product from the duplicated gene was upsetting the normal development of the fly's eyes.

Gene duplications play an important role in evolutionary change. Sometimes a mutation may inactivate a gene, creating a **pseudogene**. However, on rare occasions a mutation may alter a gene in such a manner as to give it a related, but slightly different, function. This usually occurs due to a slight alteration in the structure of the protein encoded by the gene (see "Protein Structure" in Chapter 3). Thus, an organism with a duplication followed by a beneficial mutation in one of the duplicated genes would possess two proteins with similar function. These similar genes that are located close to one another on a chromosome are often referred to as a *gene family*. The formation of a gene family by duplication is believed to be what has happened in the globin genes of humans (and primates). The globin genes exist as two gene families, one on chromosome 11 and the other on chromosome 16. Over time, they have been duplicated to form clusters of genes, and pseudogenes, that play an important role in human physiology.

As mentioned above, an unequal crossing over of the genetic information produces a duplication on one chromosome, and a deficiency in the other (see Figure 5.4). Unlike a duplication, a deficiency is rarely ever advantageous for an organism. Deficiencies result in an overall loss of genetic information and typically produce a detrimental phenotype. An example of a genetic condition caused by a deletion is cri-du-chat syndrome. Also known as "cry of the cat" syndrome, it is characterized by a cat-like cry and language problems. The condition is caused by a deletion on the short arm of chromosome 5 (5p). Individuals with the condition tend to have a shortened lifespan. Other human conditions,

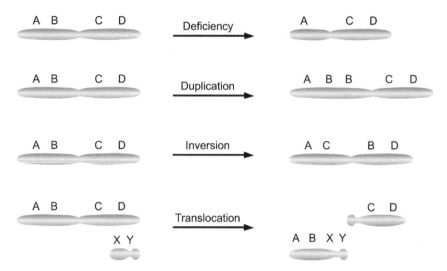

Figure 5.4 Examples of Changes in Chromosome Structure (*Courtesy of Ricochet Productions*).

such as Prader-Willi and Angelman syndrome, are due to deletions at other locations in the genome.

Another form of change in the chromosome structure is a translocation. Unlike duplications and deletions, a translocation is not due to an unequal exchange of genetic material between a chromosome pair. Instead, it is characterized as the movement of a piece of one chromosome to a new location in the genome (Figure 5.4). There are basically two forms of translocations. In a *simple translocation*, a piece from one chromosome is relocated to another position in the genome. In a *balanced translocation*, chromosomes from two different chromosomal pairs swap pieces of their chromosomal arms. Balanced translocations may not produce an abnormal phenotype, since the individual with the translocation has the correct number and types of genes, they have just been moved around the genome. However, an individual with a balanced translocation will frequently have a low fertility or greater chance of producing a child with a birth defect. This is due to the fact that during gamete formation, the translocation causes the chromosomes to align incorrectly, producing gametes that have either an excess or deficiency of genetic information.

Unlike balanced translocations, simple translocations often have a detrimental effect on the individual. This is due to the fact that an individual with a simple chromosome contains an excess of genetic information. An example is familial (or translocation) Down syndrome.

While most cases of Down syndrome are caused by an extra copy of the entire 21st chromosome (see next section on "Changes in Chromosome Number"), familial Down syndrome is caused by a translocation that moves a piece of chromosome 21 to the end of chromosome 14. An individual who inherits this hybrid chromosome 14 will have 2 copies of chromosome 14 in their cells, but three copies of some of the chromosome 21 material. The amount of material that has been translocated influences the severity of the condition. Unlike the other form of Down syndrome, familial Down syndrome is not associated with the age of the mother, and may be passed on from either parent. About 4 percent of all Down syndrome cases are familial.

The final form of chromosomal change is an inversion (Figure 5.4). In an inversion, a portion of a chromosome "flips" in its orientation. Inversions can cause problems if the break points along the chromosome occur within a gene, or if two genes become separated that have some form of positional relationship (such as gene regulation). As was the case with a balanced translocation, an individual with an inversion may appear phenotypically normal, since they possess the correct number and types of genes in their genome. Most problems with inversion occur during gamete formation. Since the chromosomes need to align with one another during the formation of gametes, a chromosome with an inversion must loop back on itself in order to align with the other member of its chromosome pair. Depending on the location and length of the chromosome, this can cause chromosomes with an incorrect number of genes, resulting in either a reduced fertility or a greater chance of genetic defects in the offspring.

These forms of chromosomal changes are rare, but overall may have an important effect on a phenotype. Each of the examples listed above can be diagnosed by a genetic counselor using a karyotype analysis (see previous section on "Observing Chromosomes").

CHANGES IN CHROMOSOME NUMBER

Most animals and plants are *diploid* organisms, meaning that they have two copies of each chromosome. Geneticists use the symbol n to indicate the chromosome number of a species. Thus a diploid organism is indicated as $2n$. For example, humans have 23 sets of chromosomes ($n = 23$), giving each nonsex cell a complement of 46 chromosomes ($2n = 46$). The term **haploid** is reserved to indicate those cells that have half the normal complement (n). In diploid organisms haploid cells are associated with sexual reproduction, as each parent contributes half of its genetic information to its offspring.

Table 5.1 Examples of Autosomal and
Sex-Chromosome Aneuploidy in Humans

Name	Chromosomal Condition
Patau Syndrome	Trisomy-13
Edward Syndrome	Trisomy-18
Down Syndrome	Trisomy-21
Klinefelter Syndrome	XXY
Jacobs Syndrome	XYY
Turner Syndrome	X0

The chromosome number in a species can vary from a few, as is the case with the fruit fly, *Drosophila melanogaster*, to several hundred, as is found in some of the insects and crustaceans. The number of chromosomes is species specific, meaning that all members of the species share the same number of chromosomes. There are some minor variations in this rule, specifically with sex determination in the insects.

Normally, chromosomes are sorted by the process of mitosis or meiosis. Mitosis occurs in **somatic cells**, while meiosis occurs in the **germ cells** (also called sex cells). The purpose of cell division is to ensure that each cell has the correct complement of chromosomes. A cell that possesses the correct complement is said to be **euploid**. However, occasionally the chromosomes do not separate correctly during anaphase of cell division. This is called **nondisjunction**. Nondisjunction causes an excess of chromosomes in one daughter cell, and a deficiency of chromosomes in the other daughter cell. If only one chromosome is involved, the condition is called **aneuploidy** and is represented algebraically as either $2n + 1$ or $2n - 1$. The term **trisomy** is used to describe the cells that are $2n+1$. They are frequently named after the chromosome that is in excess. For example, trisomy-21 represents a cell with three copies of chromosome 21. The term **monosomy** is used to represent the cell that is missing one copy of the chromosome. Thus, monosomy-13 would be a cell that only possesses one copy of chromosome 13.

In humans the impact of aneuploidy on the phenotype is dependent on whether the condition involves a sex chromosome (X or Y) or an autosome (chromosomes 1–22). Table 5.1 lists some of the aneuploid conditions in humans.

Of the autosomal aneuploid conditions, the most common is Down syndrome. Down syndrome also represents the only autosomal aneuploid condition in which the individual has a high chance of surviving past childhood. Most anueploid conditions, including Edward and Patau syndromes, have a high rate of mortality. It is believed that the majority of spontaneous abortions are due to aneuploid conditions involving an autosomal chromosome.

Down syndrome is one of the leading forms of birth defects, and occurs once in every 600 to 1,000 live births. It is estimated that there are over 350,000 individuals with Down syndrome in the United States. In rare cases, Down syndrome may be caused by a translocation of genetic material from chromosome 21 to chromosome 14 (see previous section on "Changes in Chromosome Structure"). However, most forms are caused by nondisjunction of chromosome 21. Ninety-five percent of the chromosome 21 nondisjunction events occur during the formation of gametes (eggs) in the females. As a female ages there is an increased risk of nondisjunction events. This is because female egg formation begins prior to birth. However, the process is paused before her eggs are completely formed and they remain in this dormant state until the egg begins ovulation. The longer the egg remains in the dormant condition, the greater the chance that her chromosomes will experience nondisjunction. It is estimated that women in their 20s have a 1 in 1,250 chance of delivering a Down syndrome baby. For women in their 30s, the risk rises to 1 in 400. The risk is highest for women over the age of 40 with the rate nearing 1 in 100. Individuals with Down syndrome frequently experience mental retardation and a short stature.

The major forms of sex chromosome aneuploids are listed in Table 5.1. In general, there is a greater survivability for individuals with a sex-chromosome aneuploid condition, but there are some exceptions. For example, there are not any documented cases of a male who possesses only a Y chromosome (Y0), and most X0 embryos are spontaneously aborted. Those X0 individuals who survive are said to have Turner syndrome. Turner syndrome females are characterized by extra folds of skin at the back of the neck, infertility, a high rate of hearing problems, and are often of short stature.

Another sex-chromosome aneuploid is Klinefelter syndrome. The genotype of a person with Klinefelter is XXY. These individuals are males, but do not develop full sexual characteristics and typically have low fertility. In Jacobs syndrome the genotype of the males is XYY. While studies in the past have suggested that XYY males might be prone to being more aggressive, modern research has indicated that the majority of XYY males are normal and do not display any signs of their aneuploid condition.

In addition to the production of trisomies and monosomies, nondisjunction can add an additional set of chromosomes to a cell. This is called **polyploidy** and it is the result of complete nondisjunction during cell division. The level of polyploidy is indicated as a multiple of the chromosome number of the species (n). An organism with three copies of each chromosome is called triploid ($3n$) and an organism with four copies of each chromosome is called tetraploid ($4n$).

Polyploidy is a rare event in animals, especially in the mammals. However, there are exceptions. For example, human liver cells, also known as hepatocytes, are often polyploid ($4n$ to $8n$). This is sometimes referred to as *endopolyploidy*, since only some cells within the organism are polyploidy. In *Drosophila*, an extreme case of polyploidy occurs in the cells of the salivary glands. These cells can often contain up to two hundred copies of the each chromosome. Interestingly, these chromosomes remain attached to one another, forming a structure called a *polytene chromosome*. Because of the number of copies of each chromosome, they are relatively easy to observe under a microscope, making them an ideal system for the study of changes in chromosome structure (see "Observing Chromosomes" in this chapter).

Polyploidy in plants is not only common, but it is responsible for the success of modern agriculture. Polyploidy in plants may occur in one of two different ways. First, a plant may be an autopolyploid, meaning that it contains multiples of its own chromosomes. Polyploid plants tend to produce larger fruits and flowers. In addition, many polyploid plants are sterile. This is especially true if the polyploid condition is due to an odd number of chromosomes ($3n, 5n$, etc.). Sterile plants often produce seedless fruits, which are in great demand commercially.

Alloploidy is the combination of chromosomes from two different species. An excellent example is the modern wheat plant, *Triticum aestivum* (see "Corn and Wheat: Birth of Civilization and Genetics" in Chapter 1). Wheat is an allopolyploid species that represents a combination of three separate grass genomes. Similar conditions explain the success of most of the other agricultural staples of modern society, including rice and corn. In addition, these hybrids are often more successful at surviving extreme weather conditions (drought, cold, etc.) than either of the original parental strains. This is called **hybrid vigor**. The identification of plant species that can be combined to make a more successful hybrid is a major focus of modern agriculture.

TYPES OF MUTATIONS

In the study of genetics, a **mutation** represents an inheritable change in the genetic information. A mutation may either occur in a somatic cell, in which case it will be passed on to daughter cells following cell division, or in a germline cell. Mutations in germline cells, or sex cells, may be passed on from generation to generation. For example, a mutation in a skin cell may cause a change in the phenotype of all future skin cells derived from the parent cell, but a mutation in an egg or sperm cell may influence the expression of all of the cells in an offspring.

Not all mutations are detrimental, and not all mutations result in a change in the phenotype of a cell or organism. Many mutations are silent (or neutral), meaning that they do not change the structure of the protein encoded by the gene, or the regulation of the expression of the gene. An examination of Figure 3.5 (see "The Genetic Code" in Chapter 3) reveals that the degeneracy of the genetic code allows for some mutations to be undetected. For example, a mutation that causes the DNA to change from GAA to GAG would result in a change in the mRNA codon from CUU to CUC (see "Transcription" in Chapter 3). Even though the codon has changed, the polypeptide will still contain leucine as the amino acid (see "Translation" in Chapter 3). Other mutations may occur in the introns of genes (see "Structure of a Gene" in Chapter 3) or in the intergenic regions of a chromosome. If the mutation was small it probably would not have an influence on gene expression.

A point mutation represents a change in a single nucleotide. Geneticists recognize two forms of point mutations, depending on the form of nucleotides involved. Nucleotides exist in two general forms. They are the pyrimidines (cytosine and thymine) and the purines (adenine and guanine). A transversion is a mutation that substitutes a purine for a pyrimidine, or a pyrimidine for a purine. A transition substitutes a purine for a purine, or a pyrimidine for a pyrimidine. Point mutations may be *neutral*, meaning that they have no change in the amino acid specified by the codon, or they can result in a change in the amino acid sequence of the polypeptide. These mutations are called *missense* mutations, since they have changed the meaning (sequence) of the amino acids. It is possible that a mutation causes a codon to be changed from one that codes for an amino acid to one that codes for a stop codon (see "The Genetic Code" in Chapter 3). This is called a *nonsense* mutation, since it will stop translation after that point, even if additional codons are present in the mRNA molecule being translated.

A frameshift mutation changes the sequence of codons in the process of translation. Frameshift mutations may be caused by either *insertions* or *deletions.* As their name implies, an insertion adds nucleotides to the genetic information, while a deletion removes nucleotides. Consider the hypothetical example below. In this example a "sentence" of nucleotides has been constructed:

Correct: THE FAT CAT ATE THE RED RAT
Insertion: T** HEF ATC ATA TET HER EDR AT
Deletion: TEF ATC ATA TET HER EDR AT

Notice that both the insertion and deletion change the reading frame of the message, and thus the meaning. If these were amino acids in a protein, the addition or deletion in proteins may produce a protein with altered function, or a completely dysfunctional protein. Some insertions and deletions are responsible for the formation of pseudogenes in the same manner as duplications and deficiencies in chromosome structure (see "Changes in Chromosome Structure" in this chapter).

A mutation does not necessarily have to change the sequence of nucleotides. There are a number of mutation types that result in either a structural change in the DNA sequence or a chemical modification of the nucleotides in a DNA strand. One of the more common of these is the *thymine dimer*, which is caused by UV radiation (see next section on "Agents of Mutation"). A thymine dimer is formed when UV radiation strikes adjacent thymine nucleotides, causing them to form a bond between the nitrogenous bases. Thymine dimers can cause problems during DNA replication (see "Molecular Mechanism of DNA Replication" in Chapter 3). However, most organisms have repair mechanisms to correct chemical mutations before they present a problem to the cell.

AGENTS OF MUTATION

Mutations may arise by one of two general means. First, an error in DNA replication may cause a *spontaneous mutation*. Second, a mutation may be caused by exposure to an environmental agent such as a chemical or radiation. This is called an *induced mutation*. Both induced and spontaneous mutations may be passed on to daughter cells. However, in order for either form to be inheritable the mutation must occur in a germ cell (see previous section on "Types of Mutations").

Spontaneous mutations occur as a result of an error in a cell's DNA replication machinery. Spontaneous mutations are relative because the DNA polymerase enzyme responsible for DNA replication (see "Molecular Mechanism of DNA Replication" in Chapter 3) has a proofreading function that corrects for most of the errors in replication. Overall, the DNA polymerase mismatches one in every 100,000 nucleotides. The proofreading function of the polymerase corrects all but 10 percent of these. Of this remaining 10 percent, mismatch repair systems found in many cells are capable of recognizing and repairing 90 percent. Thus, the replication machinery is responsible for an uncorrected mutation in only one nucleotide per billion.

RATES OF MUTATION

Geneticists use the term *mutation rate* to indicate the probability that a given nucleotide will be changed by a spontaneous mutation. Mutation rates are expressed as the chance of a mutation per round of replication, per gamete (in eukaryotes), or per cell division (in prokaryotes). Mutation rates also vary considerably between organisms. Bacteria and viruses tend to have higher mutation rates due to the fact that they do not generally possess efficient mechanisms of DNA repair. The mutation rate for humans is 1 $\times$ 10^{-5} to 1 $\times$ 10^{-6}. In other words, a single nucleotide has a chance of mutation once per million to 10 million gametes.

Mutation rates are not precise measurements. Rather, they are considered to represent the probability of mutation. Geneticists recognize that mutation rates are not consistent throughout the genome. Areas of the genome that contain repetitive sequences are more prone to mutation than other areas. These are called *mutational hot spots*, and they have been discovered in the genomes of most organisms.

Geneticists who are concerned with mutations at the population level use the term **mutation frequency** to indicate the rate of a specific mutation within a population. This term does not indicate the chances that the mutation will arise spontaneously within the population, but rather the estimated percent of the population that carries the mutation.

Induced mutations are caused by agents called mutagens. **Carcinogens** are a specific type of mutagen that initiates the cascade of events leading to the disease, cancer (see Chapter 7). Carcinogens usually interfere with the normal operation of the **cell cycle**. In addition, some mutagens may be classified as **teratogens**. A teratogen is a substance that causes birth defects. Some teratogens cause mutations in developmental genes, although other forms do not cause mutation they have a negative influence on the expression of developmentally important genes.

By some estimates we are exposed to over 70,000 different chemical compounds over the course of our lives. While it is difficult to determine the effects of every possible chemical combination, scientists have developed a test to examine whether a compound should be classified as a mutagen. The test is called the *Ames test* and is named after its developer, Bruce Ames (1928–). The Ames test uses strains of *Salmonella* bacteria that have been selected due to their lack of DNA repair mechanisms and ability to synthesize the amino acid histidine.

In the Ames test bacteria are mixed with the chemical of interest and then placed on nutrient plates that contain all of the essential nutrients for growth except histidine. Since the bacteria lack the ability to synthesize histidine, they will not grow on the media. However, if the chemical causes mutations, then there is a chance that some of the bacteria may regain the ability to synthesize histidine, and thus the ability to grow on the media. If the researchers know the exact number of bacteria that were in the original culture, then they can determine the **mutation rate** for the histidine gene. If this rate is above a specific threshold, then the compound may be labeled as a mutagen.

Bacteria are not mammals, and thus there existed some concern that the Ames test might not accurately portray the mutagenic effects of a compound in humans. It is also possible that the original compound is not a mutagen, but that once the compound enters into the metabolic pathways of a mammalian cell some of the intermediates may be mutagenic. Scientists have developed a variation of the Ames test in which the chemical of interest is first mixed with an extract from a mammalian liver that contains liver enzymes. This extract is then mixed with the *Salmonella* cells described above and the mutation rates calculated. Despite being around for over 30 years, the Ames test remains one of the standards for testing mutagens.

There are several methods by which a chemical can induce a mutation in the DNA strand. Some chemicals change the structure of the nucleotide. For most of these changes, the end result is that the altered nucleotide does not abide by the normal rules of complementation (see "Chargaff and Complementation" in Chapter 2). For example, ethylmethanesulfonate (EMS) adds an ethyl group to guanine, causing it to base-pair with thymine instead of cytosine. In another example, nitrous acid converts cytosine to uracil. The uracil then base-pairs with adenine, rather than the guanine that should have been paired with by the original cytosine.

Chemicals can also integrate themselves into the DNA structure and create havoc with the normal replication process. These are called intercalating agents. One of the more powerful chemicals in this class is ethidium bromide. Ethidium bromide integrates within the double helix, thus distorting the normal structure of the molecule. It is believed to be a mutagen, carcinogen, and teratogen. Interestingly, ethidium bromide has application in the study of molecular biology. After ethidium bromide integrates within the double helix it will fluoresce when exposed to UV wavelengths of light. For this reason it is frequently added to the buffer in gel electrophoresis (see Chapter 4). When the gel is exposed to UV light, the patterns of DNA in the gel are revealed. Other intercalating agents include dioxin, a group of chemicals that were used in the manufacture of the herbicide Agent Orange during the Vietnamese war.

In addition to chemicals, some forms of radiation may also cause mutation. The most damaging form of radiation are the ionizing radiations. Ionizing radiation gets its name from the fact that it ionizes, or electrically charges, a compound that it strikes by dislodging an electron from within the compound. This is a two-step process. First, the source of the ionizing radiation must release a particle (electron, helium atom) or energy. This particle then strikes a second compound, for example DNA, and dislodges an electron. Forms of ionizing radiation are classified by the type of particle that is causing the ionization. For example, alpha radiation is caused by the release of a helium nuclei from the source. Although alpha particles have a high ionizing potential, they will not penetrate very far into tissue and thus are readily stopped by the skin. Alpha radiation is not damaging to DNA unless the source is ingested. Uranium is a source of alpha radiation. Beta radiation consists primarily of electrons. It will penetrate further into tissue, but once again is not damaging to DNA unless ingested or inhaled. Carbon-14 is a source of beta particles.

The most damaging form of radiation to DNA is gamma radiation. Gamma radiation has a lower ionizing power than either alpha or beta particles, but can penetrate all of the tissues of the body. Thus, it has the greatest impact on the structure of DNA. Gamma radiation may influence DNA by one of two general methods. First, the gamma radiation may ionize a molecule in the cell and change its chemical reactivity so that it becomes a reactive with the DNA strands. An example is the **free radicals**. Alternatively, the gamma radiation may cause a break in the phosphodiester bonds between the nucleotides of the DNA strand. Often the gamma radiation contains enough energy to create a double-stranded break, in which both of the DNA strands are affected. Mutations in the nucleotide sequence then occur as the cell attempts to repair the double-stranded break. The sources of gamma radiation include cosmic rays and the radioactive decay of some elements, including radon. X-rays are also recognized as an ionizing radiation due to the high levels of energy that they contain.

Ionizing radiation is not the only form of radiation that can damage a DNA molecule. UV light, specifically UVB light, causes DNA damage by the formation of *thymine dimers*. A thymine dimer is formed between two adjacent thymine nucleotides in a DNA strand. When the UV light strikes these nucleotides, it causes a covalent bond to be formed between the bases of the nucleotides. This creates a distortion in the DNA that must be repaired before the DNA can be either replicated or transcribed. UV radiation does not have the energy necessary to penetrate deep into the tissues of the body, but it can cause problems in the cells of the skin and eyes.

Organisms have evolved a host of different mechanisms to correct damage caused by chemicals and radiation. The most common are enzyme systems to recognize thymine dimers and nucleotide mismatches in the DNA strands. These enzyme systems have a range of efficiencies depending on the needs of the organism. Overall, the complexity of DNA repair mechanisms is relatively low in the bacteria and increases in sophistication and levels in the eukaryotes. By studying DNA repair mechanisms in bacteria and yeast, researchers are able to better understand DNA repair deficiencies in humans. An example in humans of a defect in the DNA repair machinery is a condition called xeroderma pigmentosa. Xeroderma pigmentosa (XP) is an autosomal recessive condition in which the cells lack the ability to remove mismatched nucleotides from the DNA strands. Individuals with XP are very susceptible to DNA damage from UV radiation, and have a high rate of skin cancer.

In addition to enzyme systems, protection from agents that cause mutation is provided by substances in the diet, such as the antioxidants. Vitamin C (ascorbic acid) and vitamin E (tocopherol) are classified as antioxidants, as are beta-carotene (a precursor of vitamin A) and the mineral selenium. These either directly reduce the potential damage to DNA from free radicals or interact with antioxidant enzymes in the body. Fiber is known to bind many carcinogenic compounds, allowing them to pass through the digestive system without interacting with the tissues of the body.

TRANSPOSONS—THE "JUMPING GENES"

For the most part, genes are considered to be stationary objects that occupy a specific location, or locus, on a chromosome. Although genes may move locations as a result of inversions or translocations (see "Types of Mutations" in this chapter), these events are considered to be the result of chromosomal mutations, and not the action of individual genes. The possibility that sections of the genome, specifically individual genes,

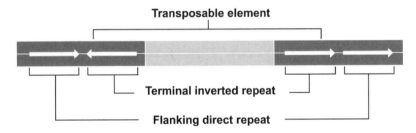

Figure 5.5 Transposon Structure (*Courtesy of Ricochet Productions*).

may move within the genome was first proposed by Barbara McClintock (1902–1992), a geneticist at Cornell University. In 1948 McClintock, published a paper describing how the movement of transposable elements, specifically the Ac and Ds elements, was responsible for chromosomal instability in maize (corn). It should be noted that this paper occurred well before the work of Watson and Crick on DNA structure (see "Watson and Crick Unveil the Double Helix" in Chapter 2). Despite not knowing the precise structure of the genetic material, McClintock was able to document through precise observations the movement of elements in the maize genome in relation to the observable phenotype of kernel pigmentation. Although McClintock reported this in 1948, her work went largely unrecognized until the early 1970s when the first mobile genetic elements were discovered in *Drosophila melanogaster*. For her work, she received the Nobel Prize in physiology or medicine in 1983.

There are a number of different types of transposons, suggesting multiple paths in their evolution. The diversity in transposons is associated with minor variations in their structure. Overall, transposons are relatively small sequences of nucleotides, consisting of several thousand base pairs. Despite their small size and diversity, there are some common structural elements that are found in most transposons. These are the *inverted repeats* and the *flanking direct repeats*. Figure 5.5 provides an illustration of how these repeats relate to the structure of the transposon. The inverted repeats mark the end of the nucleotide sequence in the transposon. As their name implies, they consist of a series of nucleotides (typically 9 to 40 bp in length) orientated in opposing directions. They are physically part of the transposon, and if the transposon moves to a new location, the inverted repeats are included in the move. Between the inverted repeats are often found the enzymes that give the transposon the ability to move, or "jump" in the genome. This movement is also called **transposition**. Key among these enzymes is *transposase*. The role of transposase is to identify the target sequence of nucleotides for insertion. In this regard the transposase enzyme has a similar function to the restriction enzymes used in molecular genetic research (see "Restriction Endonucleases" in Chapter 4). The transposase recognizes a set pattern of nucleotides, or consensus sequence, and makes an uneven double-stranded break in the DNA double helix. However, a transposon does not need to have a transposase enzyme in order to move. Since transposons are common within a genome, the transposase enzyme may be provided from another active transposon in the cell.

Following the action of transposase, the transposable element positions itself in the DNA strands. Once this has occurred, cellular repair and replication enzymes (see "Molecular Mechanism of DNA Replication" in Chapter 3) fill in the gaps around the transposon. During the repair process the flanking direct repeats are generated. Flanking direct repeats are not physically part of the transposon, but rather are a by-product of transposition. If the transposon moves to a different location in the genome the flanking direct repeats remain behind. By screening the genome for flanking direct repeats geneticists are able to examine a "history" of transposon activity.

Geneticists group transposons into three broad categories based on their method of transposition. Those transposons that move from one location to another are called nonreplicative transposons, and their method of transposition is often called "cut and paste" transposition. As these transposons move, they leave behind the flanking direct repeats. The second form of transposition is called replicative transposition, or "copy and paste" transposition. In this case the transposon remains in its original location and a copy moves to a new target sequence in the genome. Replicative transposition increases the number of total copies of the transposable element in the genome.

The third major category of transposons utilize an RNA intermediate for transposition. This RNA intermediate is generated by transcribing the transposon (see "Transcription" in Chapter 3) into an mRNA molecule. Once the mRNA molecule is generated, an enzyme called *reverse transcriptase* copies the mRNA into DNA. As its name suggests, reverse transcriptase is the reverse action of normal transcription, which copies DNA into mRNA. Reverse transcriptase is an enzyme that is not normally found in either prokaryotic or eukaryotic cells. However, the retroviruses contain this enzyme and use it as part of their infection cycle. Transposable elements that use an RNA intermediate frequently contain a gene for reverse transcriptase. The transposable elements that use reverse transcriptase are called the retrotransposons (or retroposons). Geneticists believe that the retrotransposons are very closely related to the retroviruses. Retroviruses also utilize an RNA intermediate and the reverse transcriptase enzyme. Some retrotransposons also lack the inverted repeats and direct flanking repeats common to most other forms of transposon. Retroviruses also do utilize or generate these repeats, further suggesting an evolutionary relationship between these elements.

Transposons create change in the genetic information in a variety of ways. First, as the transposon inserts itself into its new location it may disrupt either the coding sequence or regulatory regions of a gene. For example, a transposon that inserts itself into the promoter of a gene (see "Structure of a Gene" in Chapter 3) may permanently inhibit transcription, effectively turning the gene off. In addition, a transposon that jumps into the coding regions of a gene may disrupt gene expression. If the transposon leaves at some later time, the flanking direct repeats would remain behind. This introduces insertions into the gene, which very likely would cause a frameshift mutation (see "Types of Mutations" in this chapter). The frameshift mutation may inactivate the protein, and thus be detrimental, or in rare cases may provide a slightly beneficial effect by enhancing protein function. In the case where the gene is inactivated, the transposon may be responsible for the formation of a pseudogene.

Second, since transposons contain similar genetic sequences and may be present in multiple places along a chromosome, they may cause a chromosome to align incorrectly during cell division. Misalignment of the chromosome is a prime factor in the formation of insertions and deletions (see "Types of Mutations" in this chapter). Finally, the movement of a transposon into and out of the DNA strands causes double-stranded breaks in the genetic material. The formation of a double-stranded break increases the likelihood that an inversion or translocation may occur (see "Types of Mutations" in this chapter). Studies of inversions and translocations have often uncovered evidence that transposons were in the area of the break.

The human genome is littered with evidence of transposons. Geneticists estimate that between 40 and 50 percent of the genome is made up of transposon sequences. One transposon in particular, the *Alu* sequence, is found over a million times in the human genome. The sequences in these *Alu* transposons make up almost 10 percent of our DNA. There is little doubt among scientists who study evolutionary processes that transposable elements have played a major role in the evolution of life on this planet.

6

MANIPULATING THE GENE AND GENOME

The entries in the previous chapters have not only introduced the techniques used by scientists to study the gene, but also some of the diseases and conditions that are now known to be the result of changes in the genetic information. Since the discovery of DNA as the genetic material, scientists have pondered whether it would be someday possible to repair mutations or replace defective genes in a living organism. However, until recently, geneticists did not possess the technology to perform these types of experiments. But with the invention of recombinant technologies in the 1970s, it became possible for scientists to produce new combinations of genetic information and introduce it into a living cell. Then, in the next decade, technological advances led to the development of an idea that it may be possible to manipulate the genetic material while it was still in the organism. Today, many of these procedures are actively being practiced by geneticists. This chapter will explore some of these innovations and their implications on medicine and agriculture.

RECOMBINANT DNA

In genetics, the term *recombinant DNA* indicates two or more sequences of DNA that have been combined in the lab. It is important to realize that recombination occurs naturally during the process of meiosis, it is the method by which species generate new combinations of alleles. This type of recombination serves as the basis for the construction of genetic maps (see "Thomas Hunt Morgan and the Fly Lab" in Chapter 1 and "Genetic Maps" in Chapter 4), an important tool for modern geneticists. Since recombination occurs naturally in cells, each cell must contain the metabolic machinery, namely enzymes, to recombine two DNA strands.

Before scientists could manufacture recombinant DNA strands in the lab they needed to develop methods of breaking the DNA strands into smaller fragments, and then forming new DNA sequences from those fragments. In the 1970s several important breakthroughs in the study of enzymes occurred that made this possible.

The first of these was the discovery of restriction enzymes. A restriction enzyme cuts DNA at specific points based on base pair sequence. The first of these enzymes were isolated in the 1970s by Hamilton Smith and Daniel Nathans (see "Restriction Endonucleases" in Chapter 4), and geneticists quickly realized their potential for fragmenting genomic DNA into smaller pieces to facilitate study. Specifically, they recognized that some forms of restriction enzymes produced uneven cuts in the DNA. These are called "sticky ends" by geneticists (see Figure 4.1), and they are very useful in producing recombinant DNA molecules. The next discovery was the isolation of DNA ligase. DNA ligase is an enzyme that covalently links DNA strands. It has several uses in the cell, such as during DNA replication (see "Molecular Mechanism of DNA Replication" in Chapter 3) and repair. It was first isolated by H. Gobind Khorana (1922–) in the early 1970s. Together, these two discoveries gave researchers the capability of producing recombinant DNA in the lab.

The first recombinant DNA molecule was constructed in 1972 at Stanford University by two independent teams of researchers. One, headed by Paul Berg (1926–), was successful in combining a gene from a bacteriophage into a small circular viral genome. For this achievement, Berg was awarded the 1980 Nobel Prize in chemistry. The second team, consisting of Peter Lobban and Dale Kaiser, were able to piece together fragments of a bacteriophage called P22. All that was left now was to apply the procedures and introduce the molecules back into a cell.

Shortly thereafter, another team of researchers at Stanford University, led by Stanley Cohen and Herbert Boyer, recognized in 1973 that they had all of the tools with which to generate a recombinant DNA molecule, and insert it into a cell for replication. For their first system they choose the plasmid pSC101 (for more on plasmids, see "DNA Cloning" in Chapter 4). pSC101 is a relatively small plasmid, but what made it a useful tool for this experiment was the fact that it contained a single *Eco*RI restriction enzyme cut site. Thus, by treating the plasmid with *Eco*RI, it was possible to break the circular plasmid in a single location. Furthermore, *Eco*RI is a form of restriction enzyme that produces sticky ends. Sticky ends are readily recognized by the DNA ligase enzyme as places with which to restore the double-stranded nature of the DNA molecule.

For their experiment Cohen and Boyer chose to insert a gene for kanamycin resistance. Kanamycin is an antibiotic and the kanamycin resistance gene (kan^R) would allow bacterial cells to grow in the presence of kanamycin if the recombination experiment worked and the plasmid could be reintroduced into the bacterial cells. In this capacity kan^R acted as a *selectable marker*. Selectable markers remain a very important aspect of experiments involving DNA cloning (see "DNA Cloning" in Chapter 4). By cutting DNA containing the kan^R gene with *Eco*RI, they were able to generate DNA fragments that contained sticky ends. When these fragments were introduced into a solution containing *Eco*RI cut pSC101, some of the fragments were integrated into the plasmids by the DNA ligase enzyme. These plasmids could then be reintroduced back into bacteria by transformation (see "Griffith and the Transformation of DNA" in Chapter 2). When the bacteria were exposed to kanamycin, only those bacteria containing recombinant plasmid DNA would survive. These bacteria then multiplied and produced multiple copies of the plasmid and the kan^R gene. Cohen and Boyer had succeeded not only in producing a recombinant DNA molecule, but also introducing it into a cell and allowing the cell to make multiple cloned copies. With this discovery the age of recombinant technology had begun.

TRANSGENIC PLANTS AND GENETICALLY MODIFIED ORGANISMS (GMOs)

With the development of recombinant DNA techniques (see previous section on "Recombinant DNA") in the 1970s scientists quickly realized that they could use these new procedures to produce new strains of organisms that contained unique DNA sequences that could not be generated by conventional breeding strategies. A transgenic organism contains DNA from two (or more) different species. These DNA sequences are generated using recombinant techniques. In comparison, a **genetically modified organism** (GMO) does not usually contain DNA from a second species. Geneticists usually do not use this term if the organism has been changed by conventional breeding methods. Instead the term is applied when the DNA of the organism has been altered in the lab in some manner, often using recombinant technology, and then reintroduced back into the species. Transgenic organisms typically have new traits added to them, while GMOs have had an existing trait modified to reduce or improve a specific phenotype. The use of transgenics and GMOs has become widespread in agriculture, where it has been used to improve a wide variety of crop species.

The most common method of generating a transgenic plant is by using the bacterium *Agrobacterium tumefaciens*. *A. tumefaciens* is a bacterium that is found naturally in the soil. It is recognized as the causative agent of crown gall disease in a wide variety of plant species. Crown gall disease is characterized by abnormal growth of the outer surfaces of the plant, forming what looks like a tumor. These tumors are usually not lethal to an adult plant and are really only detrimental to agricultural industries that produce plants for ornamental purposes, such as nurseries.

What makes *A. tumefaciens* of interest to geneticists is the method by which the bacterium causes the formation of the tumor. When a plant is injured, *A. tumefaciens* enters through the wound and into the tissues of the plant. The bacteria contain a plasmid, called the Ti plasmid (tumor-inducing), which contains the genes for growth hormones that are responsible for the formation of the crown gall. These genes are located in a section of the plasmid that is called T DNA. After the bacteria enter into the plant, the T DNA is transferred from the plasmid into the plant cell, where it integrates into the plant genome. This movement of genes from one species to another is also known as **lateral gene transfer**.

By using recombinant DNA technology, (see previous section on "Recombinant DNA"), researchers may insert a gene of interest into the T DNA region of the Ti plasmid. Once the gene has been inserted into the plasmid, the normal infection cycle of *A. tumefaciens* introduces the gene into the plant cells where it can then be expressed. Geneticists have made improvements to the Ti plasmid by disabling genes that are associated with the formation of the crown gall, so that there is no cosmetic damage to the plant.

While the *A. tumefaciens* approach works in many plants, some species are resistant to *A. tumefaciens* infection. However, other procedures exist. One of these is called *biolistic gene transfer*, which utilizes small plastic DNA-coated microprojectiles that are blasted into the cells of a plant at a very high velocity. Another procedure uses an electrical current to the make membranes of the plant cells more permeable, allowing small fragments of DNA to enter into the cell. This is called *electroporation*. Scientists can also use microscopic needles to inject the DNA directly into the plant cell. While each of these procedures is capable of producing a transgenic plant cell, they do not have the same efficiency as the use of *A. tumefaciens*.

During the late 1980s the Calgene corporation began development of a transgenic tomato. Tomatoes present problems in both harvesting and storage, as they contain an enzyme that digests the pectins of the fruit, causing it to soften and ripen. The Calgene scientists believed that if they could disable this enzyme using recombinant techniques that the resulting fruit could be harvested easier and would have a longer shelf life. The result was the Flavr-Savr® tomato, released in 1992. Unfortunately, the tomato was not successful with the public and is no longer being produced commercially.

Another potential use for transgenic plants has been to introduce genes that provide resistance to pest species, primarily insects. One of the greatest problems with the use of conventional chemical insecticides has been the development of successful application protocols that deliver the insecticide to the plant at the height of the pest species' population size while at the same time keeping the amount of insecticide used to a minimum to reduce negative environmental effects or damage to nontarget species of animals. In the 1990s the Monsanto corporation began to work on developing a transgenic plant that contained a novel insecticide produced by a soil bacterium named *Bacillus thuringiensis*, often called simply Bt. *B. thuringiensis* produces a protein that is toxic to many species of insects, but harmless to humans. This toxin is often called Bt endotoxin. Monsanto has been successful in producing transgenic plants that contain genes for the Bt endotoxins. The endotoxins are expressed in the areas of the plant that is attacked by a pest species. Thus, only the insect pest species is impacted by the insecticide. This has been a very successful program as there are now transgenic strains of corn, potato, and soybean available. However, there have been concerns raised by the scientific community about the possibility of lateral gene transfer of the genes into other plant species and in the potential for development of resistance in the target insect species.

Monsanto has also had success in producing another group of transgenic plants that are resistant to the chemical herbicide glyphosate. Glyphosate is a powerful chemical, which is the main ingredient in the Roundup® brand of weed control chemicals. Glyphosate is toxic to most weed species, but has low toxicity to mammals. By placing a gene for glyphosate resistance into select species of plants it became possible for a farmer to spread glyphosate on an entire field since only the weed species would be affected by the chemical. Some strains of cotton and soybean now contain this modification.

These are just a few examples of the use of transgenic plants in agriculture. More are being developed and introduced into the market annually. In addition to improving resistance to pest species and the ability to harvest and store plants, scientists are also looking at ways of making crop plants less responsive to stress, or able to grow in low-water or high-salt environments. This would enable the growing of food crops on land that has historically been unable to support agriculture or has been damaged by poor agricultural practices.

GENE THERAPY

The overall goal of gene therapy is to correct an undesirable trait or disease by introducing a modified copy of the gene into a target cell. In most cases the purpose is not to replace the defective gene in the host cell, but rather to provide a new copy so that the correct protein can be expressed and the detrimental effects of the defective gene neutralized. While technically any genetic disorder may be treated by gene therapy, currently there are some limitations. First, the precise genetic mechanism of the disorder must be known, and it must be a single-gene defect. Second, scientists must know the complete genetic sequence of the gene, including regulatory regions, so that a functional copy can be delivered to the cell. Third, there needs to be an effective vector, or delivery system, for administering the correct copy to the target cells.

Generally, scientists classify forms of gene therapy as belonging to one of three types. Theoretically the most effect form of procedure would be in situ gene therapy, which means that the genetic material is administered directly to the target cells. Unfortunately it has been difficult to ensure that only target cells receive the genetic material, but there have been some successes. A second method injects the vector containing the genetic material into the fluids of the body. In this method, called *in vivo* gene therapy, the vector travels throughout the body until it reaches the target cells. A third mechanism, called *ex vivo*, removes cells from the body for gene therapy. These cells are exposed to the vector and then reintroduced back into the body. This works especially well with undifferentiated stem cells.

Scientists have several mechanisms by which the genetic information can be introduced into the target cell. The most common of which is the use of a viral vector (see Figure 6.1). Viruses are infectious agents that are frequently the choice of scientists because they typically are very specific in the types of cells that they infect. Furthermore their genomes are usually very small and well understood by scientists. The viruses that are chosen for use are derived almost exclusively from nonpathogenic strains or have been genetically engineered so that pathogenic portions of the genome have been removed. Common viral vectors are adenoviruses, retroviruses, and herpes simplex viruses. The choice of vector depends on the target and size of the gene to be replaced. In each case, after the virus infects the target cell, the DNA is either incorporated directly into the host genome, or becomes extrachromosomal.

Medical researchers are also investigating the use of nonviral vectors to deliver DNA into the target cells. As is the case with the viral vectors,

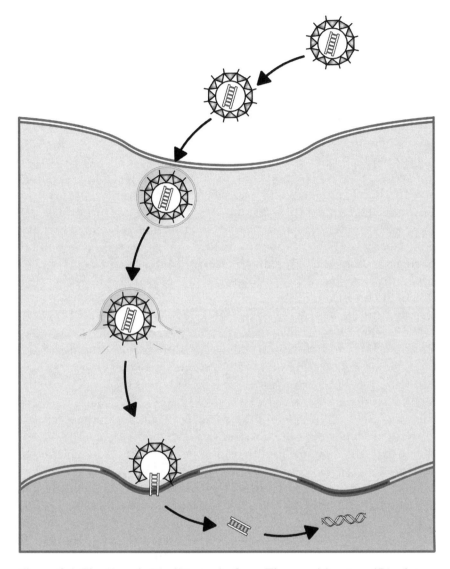

Figure 6.1 The Use of a Viral Vector in Gene Therapy (*Courtesy of Ricochet Productions*).

these mechanisms must not disrupt the normal metabolic machinery of the target cell. One system, called plasmid DNA, utilizes small circular pieces of DNA, called plasmids, to deliver the genetic material. If small enough, the plasmids can pass through the cell membrane. Although they do not integrate into the host genome in the same way as viral vectors they are a simple mechanism and lack the potential problems of a viral vector. Another mechanism being studied is to package the genetic material within a lipid-based vector, called a **liposome**, to ease transport across the membrane. However, in trials both liposomes and plasmids have displayed a low efficiency in delivering genetic material into the target cells.

Over the past decade there have been numerous scientific studies on the potential effectiveness of gene therapy to treat diseases in mammalian model species, such as the mouse and monkey. Using gene therapy researchers have demonstrated that it may be possible to treat diseases such as Parkinson's disease, sickle-cell anemia, and some forms of cancer. Gene therapy trials in humans are a relatively recent development and represent the next stage for the treatment of human diseases. Some diseases, such as Canavan disease, severe combined immunodeficiency disease (SCID), and adenosine deaminase deficiency (ADA) have already begun trials in humans. Medical researchers have suggested that in the future almost any genetic defect may be treatable using gene therapy.

While gene therapy may appear to be the silver bullet for diseases such as cancer and Parkinson's, the procedure is not without its risks. Since gene therapy using viral vectors was first proposed, scientists have recognized that there were inherent risks with the procedure. Since the technology does not yet exist to target the specific gene, chances are that the viral vector will integrate the genetic information into the genome at some site other than the location of the defective gene. This means that there is the potential for the virus to insert itself into a regulatory or structural region of a gene and in the process either render it unusable or impart a new function on the protein. Although the size of the human genome (>3 billion bases) and the fact that less than 2 percent of the genome is believed to produce functional proteins, the odds of such an event occurring are relatively low. However, given the large number of vectors used, the risk still remains a real possibility.

Two cases illustrate the risk of using viral vectors. First was the death of a gene therapy trial volunteer at the University of Pennsylvania in 1999. The volunteer, Jesse Gelsinger, suffered from a form of liver disorder called ornithine transcarbamylase deficiency (OTC). OTC is identified as being the result of a single defective gene in a five-step metabolic pathway. Using an adenovirus, researchers sought to replace the defective gene causing OTC in Gelsinger. Shortly after the gene therapy was begun, Gelsinger developed a systemic immune response to the vector and died.

The second case is actually both a story of success and failure. A French research team at the Necker Hospital for Sick Children in Paris effectively used a retrovirus vector to treat a group of young boys with SCID. SCID, also called the bubble-boy syndrome, is a rare disorder in which the immune system is rendered inoperative. One form of the disease has been traced to a gene on the X chromosome. Using the procedure of *ex vivo* gene therapy, the researchers removed stem cells from the bone marrow of the boys and, using a retrovirus vector, delivered a functional copy of the defective gene into the cells. The cells were then reinserted back into the bone marrow. The procedure was successful in that all of the boys were cured of the disease. However, 30 months later one of the boys developed leukemia, followed 4 months later by a second case. Analysis of the boy's DNA indicated that the inserted gene had disrupted a gene in which mutations had previously been shown to cause cancer.

While the number of individuals that have developed complications is relatively small, these cases do indicate the potential hazards of using a viral system and accelerated the research into using nonviral systems such as liposomes and plasmids. Additional research is underway to develop a means of targeting a specific gene. Scientists are also investigating the possibility of developing a "suicide gene," or off switch for the procedure, that could terminate treatment if an error in insertion was detected.

The science of gene therapy actually began as enzyme replacement therapy. For patients suffering from diseases in which an enzyme in a metabolic pathway is defective, enzyme replacement therapy provides temporary cure. However, since the body lacks the ability to continuously manufacture new enzymes, due to the presence of a defective gene, the therapy must be continuously administrated. In the 1980s this procedure was being used to treat a number of diseases, including ADA. ADA is a disease in which an enzyme in a biochemical pathway that converts toxins in the body to uric acid is defective. As a result the toxins accumulate, and eventually render the immune system ineffective. The modern era for gene therapy began in the early 1990s as scientists began to treat ADA with gene therapy. Through a series of trials researchers learned that *ex vivo* treatment of stem cells proved to be the most effective mechanism for treating ADA with gene therapy. In 1993, researchers obtained stem cells from the umbilical cord of three babies that were born with ADA. After the correct genes were inserted into these stem cells, the altered cells were inserted back into the donor babies. After 2 years of monitoring it appears that the process has worked and the potentially fatal effects of ADA in these children has been reversed.

Another promising area of gene therapy is the treatment of cancer. Unlike the studies mentioned above, cancer treatment using gene therapy would probably not involve replacing defective genes, but rather knocking out those genes that are causing uncontrolled cell division within cancer cells. By arresting cell division, scientists can halt the spread of the cancer. This would be especially useful in areas of the body where surgery is risky, such as brain tumors. The primary challenge at this stage is the targeting of the vector. A knockout vector would need to only infect cancer cells, and not the other dividing cells of the human body.

A potential area of gene therapy that has yet to be exploited is germ-line gene therapy. Germ cells are those that are responsible for the formation of the gametes, or the egg and sperm cells. Since these cells only contain half the genetic information of an adult cell, it is relatively easy to replace genes using available procedures learned from biotechnology. Furthermore, since following fertilization the genetic material in the germ cells is responsible for the formation of all of the remaining 100-trillion-plus cells in the human body, then any genetic change in the germ cells has the ability to be inherited by subsequent generations. Somatic cell therapy, as described above with ADA and SCID, only has the ability to influence the individual, since these cells are not normally part of the reproductive process. Gene therapy in germ cells is currently considered unethical, but many consider it to be the mechanism of eliminating certain diseases from the human species.

While currently the use of gene therapy to correct human diseases may be temporarily stalled until technical obstacles are overcome, there is little doubt in the biomedical community that gene therapy represents the procedure of the future. At the fundamental level, gene therapy has the potential to be the ultimate cure for many ailments and diseases of our species. For most of recorded history medicine has been confined to the treatment of symptoms. Since the start of the twentieth century advances have enabled enhanced surgical procedures, pharmaceutical drugs that alter or interact with the biochemistry of the cell, improved diagnostic techniques and a deeper understanding of human inheritance. Gene therapy represents the ultimate preventative procedure. Using gene therapy medical researchers can eliminate the cause of many human diseases.

KNOCK-OUT MOUSE

One application of genetics that uses the theories of both recombinant DNA and the generation of transgenic organisms is the **knock-out mouse**. The purpose of a knock-out mouse is to study the function of a specific gene by eliminating its function in all of the cells of the organism. Mice, like humans, are diploid organisms, and each of their cells contain two copies of every gene. Since it is not easy to knock out both genes simultaneously, strains of knock-out mice are generated involving a multistep process (see Figure 6.2).

In the first stage researchers begin designing a recombinant DNA molecule for a gene of interest. In one of the exons of the gene (see "Structure of a Gene" in Chapter 3) they insert a second gene. This second gene is usually *neo*, a gene that provides resistance to the antibiotic neomycin. When *neo* is placed into the exon of the gene of interest, it disrupts the coding sequence of the gene so that it can no longer produce a functional protein product (see "Transcription" and "Translation" in Chapter 3). This type of mutation is called a *loss-of-function mutation* by geneticists, since the gene has lost its ability to produce a gene product in quantities sufficient for the organism. In addition to neo, a gene called *tk* is inserted next to the gene of interest. *tk*, a gene derived from a virus, makes the cells susceptible to a chemical called gancyclovir. Both *neo* and *tk* serve as *selectable markers* (see "Cloning" in Chapter 4), which allow the researcher to check the progress of their experiment through the different stages.

Once the recombinant DNA molecule has been produced, it needs to be inserted into the cells of a mouse. For this part of the experiment, geneticists use embryonic stem (ES) cells from a line of mice that are pure-breeding for a certain coat color, usually white or agouti. The ES cells are then placed into a Petri dish containing large numbers of copies of the recombinant DNA molecules. As the DNA molecules enter into the cell, a few of them will replace the existing target gene by the process of **homologous recombination.** Homologous recombination is the replacement of a gene (or sequence of DNA) with a similar genetic sequence. Given the size of the mouse genome, this occurs at a very low rate, so very few of the ES cells will have replaced the target gene with the knockout gene. However, since the knockout gene contains the *neo* insert, these cells will be able to grow on a media that contain neomycin. It is also possible that the knockout gene could be incorporated at other sites in the genome, in which case the target gene would not be knocked out. This is called *nonhomologous recombination.* To select against nonhomologous recombinants, the ES cells are also exposed to gancyclovir. The presence of the *tk* gene, which would only be present in the nonhomologous recombinants, makes these cells susceptible to gancyclovir. The end result is ES cells that contain target genes that have been knocked out by homologous recombination.

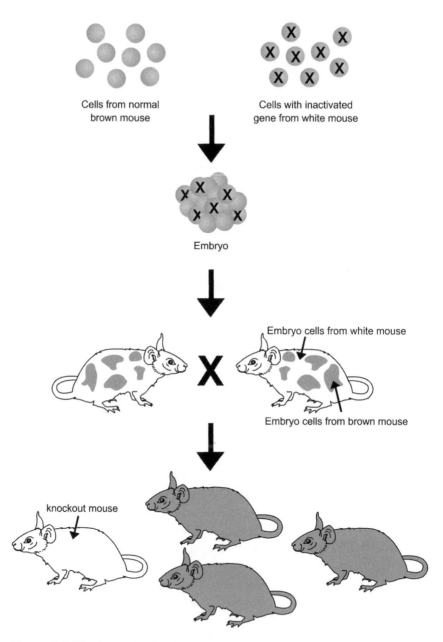

Figure 6.2 The Stages in the Generation of a Knock-out Mouse (*Courtesy of Ricochet Productions*).

In the next stage of the experiment, **blastocyst** cells are removed from a second strain of mice, usually homozygous for a black coat color. The ES cells from the white strain of mice that contain the knockout gene generated by the previous experiment are then injected into the blastocyst. The blastocysts are then placed into a female black-coat-color mouse, where they develop into new mice. The blastocyst that contains the two different strains of cells will produce a mouse that has both white and black coat color, since some of the cells originated in a white mouse (the knockout cells), while others originated in the black mouse (the blastocyst cells). These mice will have a variegated appearance. They are called **chimeras**, since they contain cells from two different sources. The chimera mice are then mated to black-coat-color mice to produce offspring that are heterozygous for the knockout gene. These mice will still contain one good copy of the target gene, while the other copy has been knocked out. If the researchers desire to produce mice that are homozygous for the knockout they can mate two of the variegated mice.

Knock-out mice provide a powerful way for researchers to study gene function. By knocking out a specific gene it is possible to determine the precise contribution of a given gene to a phenotype. This may be a physical phenotype, such as limb development, but more often than not researches use this technique to understand the impact of a gene on a metabolic pathway. Since most mice genes have homologues in humans, this procedure can provide insight into what the function of a gene is in a human. This has a tremendous application in the development of new drugs or gene therapies. In addition, geneticists have used the same procedure to develop a "knock-in" mouse. As the name implies, a knock-in mouse has had a gene added to its genome using recombination. Knock-in mice may be transgenic (see "Transgenic Plants and Genetically Modified Organisms" in this chapter) if the new gene is not from the mouse. Knock-in mice are useful in studying gene function as well. In one example, researchers were able to generate a knock-in mouse that was hypersensitive to nicotine so that they could better understand the biochemical basis of nicotine addiction.

RNA INTERFERENCE

Not all patterns of inheritance follow the laws established by Gregor Mendel in the nineteenth century. By the early twentieth century many scientists had already discovered that gene and protein interactions, and the presence of multiple-allele systems, could change the predicted outcome of a single-trait cross (see "Is Inheritance as Simple as Mendel Suggested?" in Chapter 1). By the late twentieth century most geneticists believed that the majority of mechanisms associated with non-Mendelian patterns of inheritance had been discovered. However, in the early 1990s plant geneticists were working on developing transgenic methods of increasing flower pigmentation in petunias when they noticed that instead of generating darker pigmentation from the addition of a gene, they unexpectedly produced white flower colors. The researchers believed that this non-Mendelian form of inheritance was due to "cosuppression" of the genes, although a precise mechanism was not determined. By the mid-1990s similar observations had been made in patterns of inheritance for the fungi *Neurospora crassa* and the nematode *Caenorhabditis elegans*.

In the work on *C. elegans* researchers noticed that the expression of a gene could be reduced by the addition of RNA sequences that were homologous to the gene coding sequence. At the time it was believed that these RNA sequences bound to the transcribed mRNA from the gene (see "Trancription" in Chapter 3), producing double-stranded RNA (dsRNA). It was believed that since dsRNA is not normally produced in eukaryotic cells, it was either disrupting normal translation (see "Translation" in Chapter 3) or was being targeted for destruction by **endonuclease** enzymes in the cytoplasm of the cell.

The correct explanation on how the RNA was interfering with normal gene expression was presented by Andrew Fire (1959–) and Craig Mellow (1960–) in 1998 (see Figure 6.3). As dsRNA enters into the cell, a protein called *dicer* (a type of endonuclease) chops the dsRNA into short pieces, 20–24 nucleotides in length. The dsDNA then disassociates into ssRNA (single-stranded RNA). These short ssRNA pieces then interact with a protein complex called RISC. RISC stands for *RNA-induced silencing complexes*. The RISC-ssRNA complex will then bind to any mRNA in the cell that is complementary to the ssRNA fragment complexed with RISC. Once the RISC-ssRNA complex has bound to a target mRNA, the mRNA is degraded using an endonuclease. The RISC-ssRNA complex can be reused. While the gene producing the mRNA may still be active, the RISC-ssRNA complex effectively halts expression of the gene by preventing the mRNA from being translated into a polypeptide. For their work in deciphering the mode of action of RNAi Fire and Mellow shared the 2006 Nobel Prize in physiology or medicine.

The question is often asked as to why eukaryotic cells have this type of silencing mechanism since they do not produce dsRNA. The answer is that it is a response by eukaryotic cells to the problem of dsRNA viruses. While dsRNA is not present in eukaryotic cells, many viruses utilize dsRNA as their genetic material. The virus injects the dsRNA into the cell where it then directs the manufacture of more viral particles. As

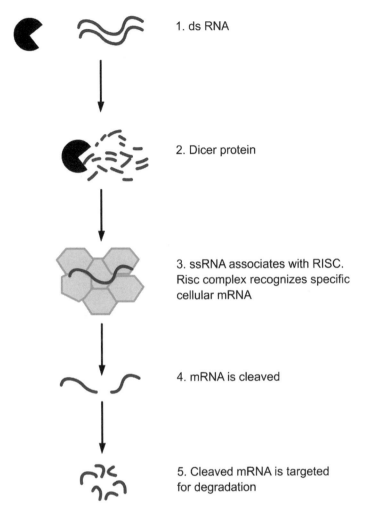

Figure 6.3 The Mechanism of RNA Interference (*Courtesy of Ricochet Productions*).

an evolutionary response, eukaryotic cells have evolved the dicer-RISC system to destroy dsRNA in the cell, thus effectively halting the viral infection. Dicer and RISC type proteins have been found in a variety of organisms, including the model organisms *C. elegans*, *Drosophila*, and *Arabidopsis*. Given the evolutionary importance of these proteins, it is believed that they will eventually be found in all eukaryotic cells.

Geneticists often refer to RNA interference (RNAi) as a "knock-down" process. The term is used because the introduction of siRNA reduces expression of the gene. In comparison a gene that has been knocked-out, as described in the entry on knock-out mice above, completely eliminates gene expression, while a gene that has been knocked in increases gene expression. The discovery of RNAi's mode of action is probably one of the most important discoveries in genetics of the past several years. It is possible for scientists to manufacture short strands of RNA with specific nucleotide sequences in the lab. These RNAs are called small interfering RNAs, or siRNA. siRNAs may then be injected into specific tissues of the body, or delivered through more systemically, in order to silence genes of interest. The applications of this technology for the treatment of diseases such as acquired immunodeficiency syndrome (AIDS) are just beginning to start. In the future it may be possible to stop infection of the immune system by the HIV (human immunodeficiency virus) by simply injecting siRNAs that disrupt viral gene expression.

SITE-DIRECTED MUTAGENESIS

Geneticists are not only interested in the structure of a gene, they also have a fundamental interest in the effects of mutation on the phenotype of an organism. The variation in phenotypes in the natural world is the result of mutations in the nucleotide sequences of genes. In order to understand the effects of mutations geneticists can wait for a gene to mutate naturally. Unfortunately, this may take a considerable amount of time given that most mutations are spontaneous and occur at a relatively low rate. The other choice is to deliberately introduce mutations into the organism. Historically, this been performed using chemical mutagens or radiation. For example, in *Drosophila melanogaster*, the mutagen ethyl methanesulfonate (EMS) is often used to introduce mutations into the genome. Since a single *D. melanogaster* female produces large number of offspring in a relatively short period of time, it is possible for researchers to produce mutations in specific genes in a short period of time. Other mutagens, such as X-rays, may be used for the same purpose (see "Agents of Mutation" in Chapter 5). Unfortunately, this approach is not possible in organisms that produce few offspring or have a long generation time. Even in mice geneticists would have to perform numerous exposures to EMS or X-rays in order to generate a single mutation of interest.

With the invention of DNA sequencing and PCR technologies (see Chapter 4 for more information), it became possible for researchers to more quickly produce specific mutations within specific sequences of nucleotides. This process is called *site-directed mutagenesis*. In order to perform site-directed mutagenesis the researcher must have some prior knowledge of the target gene. Usually the target gene has been cloned and sequenced already and the researcher simply wants to understand the influence of a specific mutation on gene function. Using chemical synthesis procedures, the researcher generates a short sequence of nucleotides, usually 20–22 bases in length, that corresponds to a nucleotide sequence within the target gene. This stretch of nucleotides is called an *oligonucleotide.* The only difference between the oligonucleotide and the sequence of the target gene is the fact that the researcher introduces a single-base pair substitution into the oligonucleotide.

In the next step of the experiment the oligonucleotide is placed with a ssDNA vector, usually from a virus. The vector contains a cloned copy of the target gene (see "Cloning" in Chapter 4). The oligonucleotide sequence will match up to the complementary region of the vector, producing a small segment of dsDNA. Once this has occurred DNA polymerase, nucleotides, and DNA ligase are added to the mixture. The DNA polymerase will then replicate the missing sequences by the process of DNA replication (see "Molecular Mechanism of DNA Replication" in Chapter 3), making the entire vector double-stranded. The strands will be complementary with the exception of the one spot where the mutation was introduced into the DNA sequence. The vector in then placed within a living cell. Once there the DNA repair mechanisms recognize that there is a mismatch between the two strands. The repair mechanism has one of two choices, it can either repair the strand using the sequence in the oligonucleotide as the template, or it can use the original gene sequence as the template. If the oligonucleotide is used as the template, both copies of the gene will now contain the mutation.

Once generated, the researcher can then introduce the sequence back into an organism to view the effects of the mutation on a phenotype. For mammals, these altered genes can be used to generate knock-out mice (see "Knock-out Mouse" in this chapter) and in plants the process can be used to produce a genetically modified or transgenic plant (see "Transgenic Plants and Genetically Modified Organisms" in this chapter for more information). This is especially beneficial in the understanding of gene function and in the study of diseases that are due to single-base mutations in a gene, such as Parkinson's and some forms of Alzheimer's disease.

7

APPLICATIONS OF GENETICS

Despite the advances in our understanding of genetics, geneticists are still a long way from knowing everything about how our genome functions. New discoveries are being made every day. With each new discovery usually come new questions and the development of new techniques. Geneticists are also finding new ways to apply their knowledge toward solving some of the problems of the human race, from crime to complex diseases. This chapter will explore a few of the advances in modern genetics and provide some insight into how geneticists study the very complex problems of behavior and human disease.

POPULATION GENETICS

Scientists define a population as a group of individuals of the same species occupying a given geographic area. A population geneticist studies how the frequency of the alleles changes in a population over time, usually in response to some selective force. In this regard, population geneticists are usually less interested in the study of molecular mechanisms of gene regulation, but rather use mathematics and statistics in order to describe the changes in a population. For example, as the basis of their science, population geneticists use a mathematical formula called the **Hardy–Weinberg equation**.

The Hardy–Weinberg equation is also called the *Hardy–Weinberg equilibrium* and the *Hardy–Weinberg law*. It is named after the work of two early twentieth-century mathematicians, Godfry H. Hardy (1877–1947) and Wilhelm Weinberg (1862–1937). Working from information provided earlier in the century by the American mathematician William Castle (1867–1962), Hardy and Weinberg were able to independently derive (ca. 1908) that the algebraic formula

$$p^2 + 2pq + q^2 = 1$$

could be used to describe how the allele frequencies in a population would respond under a given set of conditions. This formula is used for the study of a single gene with two alleles. The symbol p represents the allele frequency of one of the alleles in the population, while the q represents the allele frequency of the other allele of the gene. The frequency of p and q in a population are obtained by extracting DNA from a tissue sample (for example, blood) and then amplifying a specific gene using PCR (see PCR, in Chapter 4). The individual alleles of the gene can be identified by looking for any number of molecular markers within the gene (see "Molecular Markers" in Chapter 4), including RFLPs and SNPs. Once the alleles have been identified, the number of alleles in the population can be determined. This is called the **gene pool**. Since many organisms are diploid, each individual contributes two alleles to the gene pool.

The variables in the Hardy–Weinberg equation reflect the possible genotypes of an individual. For example, suppose in the gene we are examining there are two alleles, B and b. If the frequency of allele B is 60 percent in the population, and the frequency of b is 40 percent in the population, then individuals who are homozygous for allele B (BB) will constitute 36 percent of the population ($0.60 \times 0.60 = 0.36$ or 36%). Since p represents the frequency of B in the population, then p^2 is the percent of individuals who are BB. Similarly, q^2 represents the percent that are homozygous recessive (bb) and $2pq$ represents the percent that are heterozygous. What the Hardy–Weinberg equation establishes is that if a population is stable, then the allele frequencies of p and q should not change from one generation to another. A population that is stable is said to be at equilibrium.

Several conditions must be met in order to be considered a stable population or one where the frequency of the alleles in the population are not changing over time. First, the population size being studied must be large. Often population geneticists are not able to sample an entire population, but rather take a representative sample and then predict the allele frequencies. If the population size is small, then there is a chance that random sampling error may have an adverse effect on the initial prediction of allele frequencies. Second, there must be random mating in the population. Remember that population geneticists are not interested in individuals, but in the frequency of alleles in the gene pool.

So in order for the frequency of alleles to be at equilibrium in the gene pool, all of the alleles must have an equal chance of being introduced into the gene pool. Nonrandom mating means that some form of sexual selection is occurring, which influences the chances of an allele entering in the next generation's gene pool. The third condition is that there cannot be any mutation. Causes of mutation are discussed in Chapter 5. Although the chances of a mutation in the specific gene being studied during the course of the study is low, mutation does have the potential of introducing new alleles into the population. The introduction of new alleles immediately moves a population out of equilibrium. The fourth condition is that there cannot be any migration into (immigration) or out of (emigration) the population. Different populations may have different frequencies of the same alleles, and any migration between the two populations would cause the stable population to leave equilibrium. The final condition is that there can't be any selective pressure against the gene being studied. Natural selection is a powerful force, which favors beneficial alleles and acts against detrimental alleles. Selection for or against an allele will move a population out of equilibrium.

After a closer look at these conditions it should become apparent that stable populations are rare in the natural world. In fact, that is the overall reason why population geneticists use the formula. Having established what the alleles' frequencies should be if the population was in equilibrium, population geneticists can measure the effects of outside influences on the population. For example, selection against a certain allele will cause a population to go out of equilibrium so that the allele frequencies from one generation to the next are not stable.

Conservation geneticists routinely use the principles of the Hardy-Weinberg equation. One of the greatest concerns of conservation genetics is the principle of **genetic drift**. Genetic drift is the influence of chance events on the frequency of alleles in the gene pool of a population. Theoretical geneticists, such as the influential statistical geneticist Sewall Wright (1889–1988), have devised formulas to predict how chance events may influence allele frequencies. Those formulas are beyond the scope of this book, but basically they state that over time, genetic drift serves either the fixation or loss of an allele from a population. Furthermore, the smaller the population is, the faster the influence of genetic drift. Since conservation geneticists often work with small populations of endangered animals and plants, they are frequently concerned that genetic drift may fix a detrimental allele in the population.

Two examples of genetic drift are the *bottleneck effect* and the *founder effect*. Bottleneck effects are sudden events that drastically reduce the population size. Natural disasters are excellent examples of a bottleneck effect. The reason why this is considered a chance event is because the survivors of the disaster are usually not selected for based upon their genotype, but rather have survived due to chance. Bottleneck effects can drastically chance the composition of the gene pool since the gene pool of the survivors is usually vastly different from the original population. A founder effect has similar consequences. In the founder effect a single or small group of females is separated from the original population and establishes a new population. This new population needs to be isolated enough from the original population so that there is no migration between the two populations. If this occurs, the new population may contain a very different frequency of alleles in its gene pool than the original population. A nice example of this would be the movement of birds from a mainland area onto an isolated island far out to sea. This is believed to have been the case for many of the bird species found on the Galapagos Islands in the Pacific Ocean. As was the case with the bottleneck effect, it is usually chance that causes a founder effect, not selection against or for a specific allele in the gene pool.

Population genetics also forms the basis of the study of **microevolution**. Microevolution is the study of changes in the allele frequencies of a population over time. Most often this involves the influence of a selective force over relatively short periods of time. For example, the use of an insecticide on a pest insect species will reduce the number or percent of insecticide-susceptible alleles in the population, but will increase the frequency of resistant alleles in the population. This can usually occur within one to two generations. In comparison, **macroevolution** is the study of large-scale trends in evolutionary events over long periods of time, often millions or billions of years.

Human population geneticists who are interested in the movement of human populations across the globe often use mitochondrial DNA (mtDNA) as the basis of their population studies. The use of mtDNA has several benefits over nuclear DNA. First, mtDNA has a higher rate of mutation than nuclear DNA. This is primarily due to a lack of DNA repair enzymes. Therefore there is potentially more variability that can be used as a form of molecular marker (see "Molecular Markers" in Chapter 4). Second, mtDNA is inherited solely from the mother, therefore all of the mtDNA in each of our cells is the same as our mother's mtDNA. This provides an unbroken line from mother to offspring. By studying patterns of sequence variation in the mtDNA researchers are able to trace the movement of groups of humans around the globe and establish the historical origins of isolated groups of humans in remote areas of the world.

In addition to the areas of study mentioned above, the principles of population genetics can be applied toward understanding human disease and the potential influences of epidemics and pandemics on the human population. Furthermore, the principles of population genetics, specifically the establishment of allele frequencies in groups of humans from different areas of the globe, have been applied in the development of DNA fingerprinting, a useful tool in the study of forensics and anthropology (see next section on "DNA Fingerprinting").

DNA FINGERPRINTING

Fingerprinting has been a potential source of identification for over a century. Since fingerprints are unique, they are often used as a source of identification in criminal cases. However, fingerprints are not permanent and are dependent on the type of surface that the individual contacts with their fingers. It is also relatively easy to avoid leaving fingerprint evidence at the scene of a crime by simply wearing gloves or wiping surfaces.

The sequence of nucleotides in our DNA represents a unique form of fingerprint. With the exception of **monozygotic twins**, no two individuals have the same patterns of nucleotides in their genomes. Therefore, technically if even a single cell is left at the scene of a crime investigators can establish that an individual was present at the location. While sometimes controversial, DNA evidence has proven to be a powerful tool in establishing both the guilt and innocence of a suspect.

The concept of DNA fingerprinting, also known as *DNA profiling*, was first introduced in the mid-1980s by Alec Jeffries based on his study of repeated sequences in the genome called *minisatellites*. Minisatellites are a form of molecular marker (see "Molecular Markers" in Chapter 4) that vary in the number of times that a sequence of nucleotides is repeated in tandem. They belong to a class of markers called the *variable number of tandem repeats*, or VNTRs. Initially DNA fingerprinting was performed using Southern blots, restriction enzymes, and DNA probes (see Chapter 4 for more information on these procedures) to generate the results. However, with the invention of the PCR in the 1980s (see "Polymerase Chain Reaction" in Chapter 4) it became possible to amplify the areas containing the molecular markers directly, thus greatly accelerating the process.

In the PCR procedure, researchers first amplify a region of the DNA that contains the minisatellite marker. The length of the DNA fragments produced by the PCR reaction will depend on the number of repeats that exist within the amplified region of DNA. A minisatellite that contains just a few repeats will be shorter than a minisatellite containing a larger

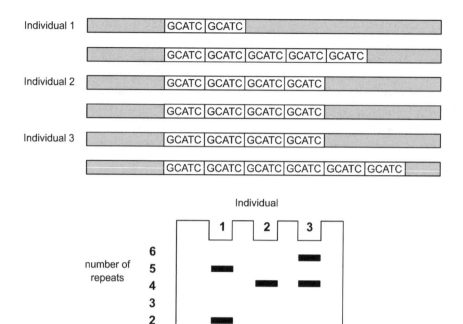

Figure 7.1 An Example of a DNA Fingerprint (*Courtesy of Ricochet Productions*).

number of repeats. The shorter DNA fragment will migrate faster on an agarose gel (see "Gel Electrophoresis" in Chapter 4). Since each person has two copies of the minisatellite (one copy from each parent), they should produce a two-band pattern on the gel. Occasionally an individual may have inherited minisatellites with the same number of repeats from both parents, in which case there would only be a single band on the gel. These patterns can then be compared against the evidence (see Figure 7.1).

A single set of amplified bands on a gel is not going to provide conclusive evidence for or against an individual in a crime case. Therefore, geneticists that work with DNA fingerprinting often amplify multiple (5–7) minisatellite regions of the genome. This provides a much more definitive DNA profile to examine. Geneticists are also now using microsatellites, or short tandem repeats (STRs) in their work. The science of population genetics (see previous section on "Population Genetics") has also been applied to the use of micro- and minisatellites in DNA fingerprinting. Using reference populations, population geneticists have determined the relative frequency of each type of minisatellite allele that occurs in the population. For example, if minisatellites with length

A occurs in 13 percent of the population, and minisatellites of length B occurs in 5 percent of the population, the odds of getting both A and B are 0.65 percent. The more markers that are analyzed, the more likely it is that the fingerprint is accurate.

DNA fingerprinting was first used in 1987 to convict a rapist in England and is now routinely used in criminal investigations. However, it does have other applications as well. Conservation geneticists can use DNA fingerprinting to identify the source population of a poached animal, and agricultural geneticists can use the procedure to identify specific strains of plants. Some people believe that within a few years DNA fingerprints will replace traditional forms of identification such as Social Security numbers and driver's licenses.

CANCER GENETICS

Cancer is the result of mutations that change the way that a cell behaves in a tissue. Basically, cancer represents a group of cells that are no longer following their genetic programming. In order to understand the genetics of cancer it is first necessary to take a look at the cell cycle. The cells of the body that undergo normal cell division are called *somatic cells*. As they progress toward cell division, each somatic cell passes through a series of stages which together are called the cell cycle. Cell biologists have defined the stages of the cell cycle based on what is going on within the cell in order to prepare for cell division. The basic stages of the cell cycle are illustrated in Figure 7.2. The two most important phases of the cell cycle are S phase and the process of mitosis and cytokinesis. During S phase the cell replicates its DNA (see "Molecular Mechanisms of DNA Replication" in Chapter 3) so that it can pass an exact copy of its genetic material onto the next generation of cells. During the process of mitosis and cytokinesis the cell actively divides, forming two new daughter cells.

As the cell progresses through the cell cycle it must pass a series of checkpoints. At each of these checkpoints a series of proteins chemically determine whether the cell is prepared to pass onto the next stage of the cell cycle. The two major checkpoints are the G1/S checkpoint and the G2/M checkpoint (see Figure 7.2). These proteins are under the influence of a variety of environmental factors, including growth factors and nutrition. Mutations in the genes that encode for the checkpoint proteins may cause the cell to enter into unrestricted cell growth, a key characteristic of cancer.

These genes fall into two general classes. The first are the *tumor-suppressor genes*. When functioning normally, tumor-suppressor genes

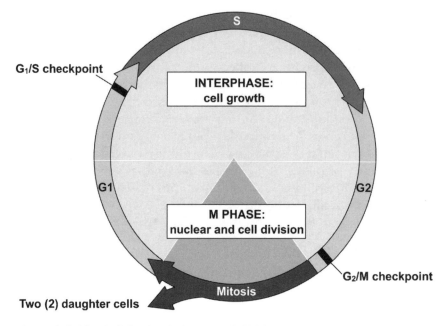

G₁/S checkpoint

INTERPHASE:
cell growth

G1

G2

M PHASE:
nuclear and cell division

G₂/M checkpoint

Mitosis

Two (2) daughter cells

Figure 7.2 The Cell Cycle of a Somatic Cell (*Courtesy of Ricochet Productions*).

serve to stop, or slow, cell division by preventing the cell from entering into the next cycle. Opposite of these are the *proto-oncogenes.* Proto-oncogenes are those that promote cell division. It is important to distinguish a proto-oncogene from an oncogene. An oncogene is one that gives the cell the ability to proceed through the checkpoints of the cell cycle without regulation. In order to become an oncogene, a proto-oncogene must obtain a mutation that changes its overall function in the cell. Cells do not contain oncogenes, they contain proto-oncogenes that when mutated result in unregulated oncogenes. Since humans contain two copies of each gene, it would seem initially that a mutation in one of the copies of the gene should not cause cancer since an unmutated copy of the gene still exists in the genome. This is true for mutations involving tumor-suppressor genes. In order for a cell to lose its tumor-suppressing function both alleles must be inactivated by mutation. This is sometimes called the **two-hit hypothesis**, a concept of cancer genetics first proposed by Alfred Knudson (1922–) in the 1970s. However, the two-hit hypothesis does not usually apply to the proto-oncogenes. Mutations in these genes, if associated with cancer, cause the gene to obtain the new function of increasing cell division. Thus, only a single mutation is necessary in either of the two copies of the gene. An example of a tumor-suppressor gene in humans is the *RB* gene. When mutated this gene is responsible for the development of a type of cancer called retinoblastoma which is characterized by tumors in the eyes. Another example is *BRAC1*, a tumor-suppressor gene that is associated with breast and prostate cancer. An example of a proto-oncogene is *ras*. This gene is associated with a number of cellular signaling pathways that regulate the cell cycle. When mutated, the resulting proteins allow the cell to ignore the checkpoint signals, leading to the formation of a wide variety of cancers.

The mutations that inactivate the tumor-suppressor genes, or change the function of the proto-oncogenes, can be caused by a variety of factors. Many of these are covered in the "Agents of Mutation" entry in Chapter 5. Cancer is usually not inherited, although some individuals may possess genotypes that increase their susceptibility for the development of cancer. In order for a cancer to be inherited there would need to be a mutation in the *germ cells* of the body. Most cancers are sporadic, meaning that they occur in the *somatic cells* of the body. However, sporadic cancers may spread to other parts of the body in a process called *metastasis*.

Cell biologists and geneticists have identified several key characteristics of cancer cells. First, the mutations that cause a cell to become a cancer cell often causes the cells to be *undifferentiated*. In general, the somatic cells of the body are highly specialized for a unique task. They have a defined shape and cellular characteristics. Shortly after a cancer mutation occurs in the cell the cell changes shape. It develops an irregular shape that easily distinguishes it from other cells. During a biopsy a physician takes a sample of the tissue that is believed to contain cancer cells and looks for irregular shapes under the microscope. In addition, cancer cells are often immortal, a characteristic that is of intense interest to scientists who study both cancer genetics and the genetics of aging. Most cells undergo programmed cell death, or *apoptosis*, at some point. Cancer cells are able to bypass this control indefinitely by activation of an enzyme called *telomerase*. For example, a strain of cancer cells called HeLa cells were derived from the cervical cancer cells of Henrietta Lacks (d.1951) in the 1950s. These cells are now routinely used in the study of cell biology and genetics throughout the world. The telomeres of each chromosome act as form of cap (see "Chromosome Structure" in Chapter 5). With each cycle of DNA replication the telomeres shorten. When they reach a critical length the chromosome becomes unstable and enters into apoptosis. Some cells, namely stem cells, possess an enzyme called telomerase that renews the telomeres following each cell division, thus allowing the cell to escape apoptosis. Most cancer cells also possess an active telomerase enzyme, effectively making them immortal.

Since cancer will effect one in three individuals over the course of their lives, it remains one of the areas of genetics that many people will have personal experience with some time in the future. Advances in screening, such as the use of microarrays (see "DNA Microarrays" in Chapter 4), and advances in treatment have caused a reduction in mortality for many forms of cancer. However, geneticists are still working to discover the causes of many forms of cancer.

BEHAVIORAL GENETICS

In the simplest sense, *behavior* is defined as the way in which an organism responds to its environment. Behavior is found in all organisms, since one of the characteristics of life is the ability to respond to stimuli. In animals, behavior is under the control of the nervous system. Regardless of the level of animal complexity, all nervous systems function in fundamentally the same manner. A network of cells must perceive stimuli, integrate incoming information, and deliver a response action. This is all made possible by the use of chemical signaling molecules called *neurotransmitters*. Within the genome of every animal species (with the possible exception of the sponges) are genes that direct the production of neurotransmitters. Unlike many of the other genes discussed in Mendelian genetics, which exist in a simple on/off configuration, the neurotransmitter genes are finely regulated so that the levels of chemicals in the synapses of the neurons can be precisely controlled. In addition to neurotransmitter genes, there are also coding sequences that dictate the structure of the specialized receptors that respond to the presence of the neurotransmitters. For each cell in the organism, genetic controls indicate the presence and number of neurotransmitter receptors on the surface of the cell, thus adding an additional level of control for the response. Simply put, nervous systems, and their genetic controls, evolved so that the organism could dictate the precise response of any cell within the body to a wide range of stimuli. Despite this level of complexity, no organism can possibly anticipate every form of environmental stimuli that it will encounter during its lifetime. In other words, the genome of a species does not possess a dedicated gene for every variant of an environmental stimulus. Rather, the nervous system of complex organisms represents a vast interaction of structural and regulatory genes.

We now know that most behavioral traits are not the result of single genes (although we will examine a few examples), but rather represent an interaction of **polygenic** mechanisms with environmental factors. The interaction of a gene with the environment produces a **multifactorial** trait. The multifactorial basis of many behavioral traits adds a new level of difficulty for geneticists, since experimental procedures must be designed to assess the influence of the environment on the behavioral phenotype. To facilitate studies of multifactorial behavioral traits geneticists frequently use model organisms, such as the mouse and the fruit fly, in their studies. By using model organisms, scientists can control specific variables, such as environmental input, that may not be possible in studies utilizing humans.

NURTURE VERSUS NATURE

One of the big questions facing behavioral geneticists is whether our behaviors are genetically hardwired into our nervous systems before we are born, or is behavior based exclusively on the environment to which we are exposed to during our lifetime? This question represents one of the great debates not only in the biological sciences, but also in the fields of sociology and psychology. It is commonly called the "nurture versus nature" controversy, and it has its beginnings in the writings of Hippocrates in the fourth century BCE In this century, geneticists have tended to favor the influence of genes in the modeling of behavior, while psychologists have stressed that the environment (nature) plays the key role in the shaping of human behavior.

Despite the apparent polarity of psychological and genetic positions, most researchers now recognize that both the environment and genes, in varying degrees, are influential in human behavior. Modern studies have identified three different relationships between genetics and the environment that can be used as models in understanding behavior. The *genotype-environment interaction* model suggests that certain genotypes have a susceptibility to environmental influence. Opposing this idea is a more recent model that proposes the concept that individuals create their own environments based upon their genetic heritage. Geneticists call this the *genotype-environment correlation*, but in the common literature it is also known as the *nature of nurture*. It represents a genetic control over environmental exposure, the details of which are still being worked out by researchers. Finally, there is the model of *nonshared environment*. Studies of nonshared environment attempt to explain why the environmental influences shared by a family (i.e., family life) are less influential on a behavioral phenotype than the shared genotype of the family.

Given the complexity of these models, how does a behavioral geneticist distinguish between environmental and genetic influences? The use of model organisms such as *Drosophila*, make is possible to control environmental variables and examine the effects of specific gene mutations on a behavioral phenotype. Conversely, genetically identical organisms can be exposed to a variable stimulus and the effect on behavior monitored.

Such approaches have been used very effectively in the study of behavior in *Drosophila melanogaster* and the mouse.

In humans the study of twins and adoption provides an excellent opportunity to distinguish the influence of genetics and the environment on behavior. There are two classes of human twins, those that arose from a single fertilization event, and those that are formed from the fertilization of multiple eggs. The first are called *monozygotic twins*. Monozygotic twins are genotypically, and usually phenotypically, the same. They are often mistakenly called identical twins, however, small developmental changes can produce differences in their phenotype. If the twins are raised together throughout their childhood, then the environment is considered to be a constant. However, if they are separated at birth, as is the case with adoption, then the effects of two unique environments on a single genotype can be examined.

The second class of twins are called *dizygotic twins*, also known as fraternal twins. Dizygotic twins occur when multiple eggs are fertilized. Usually, each embryo possesses its own placenta and amnionic sac. Since each represents a separate fertilization event, they share only 50 percent of their genes with one another, the same percentage as any two siblings. They are genotypically and phenotypically different. However, since they are raised together they allow researchers to examine the simultaneous influence of environmental factors on two different genotypes. It is important to recognize that the comparisons of monozygotic and dizygotic twins, combined with adoption studies, provide a natural experimental system for behavioral researchers to examine whether a specific behavior is the result of nurture or nature.

———————————————— ✐✐✐ ————————————————

An excellent example of a behavioral trait that can easily be quantified occurs in the maternal care behavior of the mouse. Researchers have identified a gene in the mouse genome that is active in the brain that is responsible for the "motherly" instincts of a female mouse. The gene, *fosB*, acts as an activation mechanism for the maternal behavior pathways in the hypothalamus of the mouse brain. Following birth, the sounds, smells, and touch of the newborn pups serve to turn on *fosB*, which in turn activate the maternal care pathways. The new mothers will now spend time with their offspring, nurse them, and retrieve them if they stray from her side. However, if these genes are inactivated (see "Knockout Mouse" in Chapter 6), the mother does not develop the maternal behavior and loses interest in her offspring. Many behaviorists use this as an example of nature (a genotype) regulating nurture (a behavior). *fosb* genes and gene products (namely delta-fosB) have been shown to be involved in other behaviors as well, including addiction to chemicals such as cocaine.

Another example of a behavior in a model organism that is worth examining is the genetics of mating behavior in *Drosophila melanogaster*. Like many animals, the mating ritual in *Drosophila* is a complex behavior that involves a series of courtship steps prior to physical mating. These courtship steps are evolutionarily important in that they serve to identify that the two individuals are part of the same species, as well as establish the fitness of the males. As such, courtship rituals are under intense selective pressure. In most cases in the animal kingdom, including *Drosophila*, the female acts as the selective agent during mating. In other words, it is the male's responsibility to perform a species-specific dance that displays his fitness to mate. The male courtship behavior of several *Drosophila* species is well documented, and in the case of *Drosophila melanogaster*, several genes have been identified that are associated with courtship and mating behavior. What is interesting is that each of these genes appears to have multiple roles in the organism's physiology, in other words, the courtship genes of *Drosophila*, and possibly in other animals, are **pleiotrophic**.

The courtship behavior of *D. melanogaster* is illustrated in Figure 7.3. The ritual involves visual, vibration, and chemical (pheromone) signals. To begin the mating ritual the male first orientates himself toward the female and then taps her on the abdomen with one of his forelegs. It is now known that both male and female fruit flies release species-specific pheromones that assist in species identification. These chemicals exist as complex hydrocarbons on the cuticle of the insect. They appear not to be involved in long-range attraction, as they are not very volatile, but probably play a role in the first few initial stages of mating. If the female does not move away from the approaching male, the male fly will extend his wing(s) and produce a species-specific "song." The song,

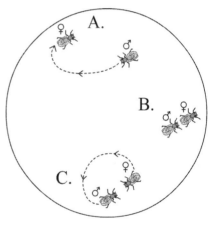

Figure 7.3 Drosophila Mating Behavior (*Courtesy of Ricochet Productions*).

which is actually wing vibrations, contains two distinct components, a sine and a pulse. As the male performs the song he will move back and forth in front of the female in an arc-like motion. He will periodically swing behind the female, and then back in front while conducting his song. During this ritual the female is usually covering her genitals with her wings, if she is interested in the male, she will move her wings and the male will attempt to lick her genitalia using his proboscis. If the female remains receptive, he will attempt copulation. If she is not receptive, the male will repeat the procedure.

In the 1970s a number of researchers identified that the courtship ritual of *Drosophila* involved the use of multiple areas of the central nervous system, thus it was recognized that multiple genes were probably involved in the behavior of courtship. Additional candidate genes are being discovered on a regular basis and research is expanding into studies of other *Drosophila* species, namely the sibling species *Drosophila simulans.*

The *period* gene (*per*) is associated with the circadian rhythm, or internal clock, of fruit flies. It is believed to have homologues in all animals. The circadian rhythm of an organism determines many functions, including sleep-wake behavior. Mutations in *period* were shown to lengthen and shorten the internal timekeeping of the organism. Other genes, such as *timeless,* had a similar effect on the circadian rhythm of the organism. In the early 1980s, researchers began to wonder if *period* may also be associated with other behaviors that required a regular timekeeping mechanism. As noted, the song produced by the male during courtship is species specific, and consists of precisely timed pulses of wing activity. Variations in the song would not be favored by sexual selection. The researchers found that male flies with an inactive *period* gene demonstrated a reduced ability to court females, and thus a reduced fitness. Thus, *period* is an interesting gene in that its gene products have a regulatory effect on two unrelated behaviors, the sleep-wake cycle and courtship.

The *fruitless* gene (*fru*) is another example of a behavior gene that has pleiotropic effects on the organism. Male flies with an inactive *fruitless* gene may display multiple changes in the mating ritual. The males are unable to distinguish between male and female flies, and will actively court either sex. Male flies with a defective *fruitless* do not produce a correct song, and alleles of the gene completely inhibit any form of copulation. The *fruitless* gene product is expressed in the central nervous system of flies, and is active during the pupal period of the fly's development. It is believed that during this stage of development the male sexual behavior of the fly is programmed into the central nervous system.

Male fruit flies also have the ability to learn during the courtship ritual and can respond to the female's actions. A number of genes have been shown to be important in the learning behavior as it relates to courtship. Mutations in the *rutabaga, dunce,* and *Cam*KII all upset the learning behavior of flies and disrupt subsequent attempts at mating. It

Table 7.1 Behavioral Traits in Humans that May Be under Genetic Control

Emotional stability	Stress response
Social withdrawal	Self-control
Dementias	Biopolar
Autism	Anxiety
Schizophrenia	Depression
Temperament	Fearfulness
Memory	Intelligence
Extroversion	Introversion

is interesting to note that to date there has not yet been a single gene identified in *Drosophila* that is dedicated solely to mating behavior. This serves to demonstrate that mating behavior in *Drosophila*, and possibly all animals, is a complex behavior with multiple, seemingly unrelated, gene products interacting to produce the correct phenotype.

From their success with model organisms behavioral geneticists have begun to examine the genetic basis of human behaviors. A number of human behaviors have demonstrated some level of **heritability**. These are indicated in Table 7.1. However, to date geneticists have not found a human behavior that is under strict genetic control, indicating that human behavior is a complex mix of environmental and genetic, most likely polygenic, factors.

Bipolar disorder is a disease of the central nervous system and is the result of an imbalance of serotonin, a neurotransmitter found in the brain. People with bipolar disorder have intense bouts of depression, sometimes lasting weeks, followed by brief periods of mania in which the person exhibits abnormal overactive behaviors. One of the problems with diagnosing bipolar disorder is that it is frequently confused with the similar phenotypes of common depression or major depression disorder (MDD). However, twin studies indicate that biopolar disorder has a higher level of genetic control than MDD.

The genetic study of bipolar disorder has focused on possible linkage to the mitochondrial genome, as well as chromosomes 4, 12, 16, 18, and 21. Studies have identified candidate genes involved with serotonin transporters (the gene *hSERT*), serotonin receptors, and a neurotransmitter called monoamine oxidase. The action of this transporter is the target of selective serotonin reuptake inhibitor drugs such as fluoxetine (Prozac). However, none of these candidates can explain all of the forms of the disease, which suggests that bipolar disorder may be a complex trait that is based on genetic **heterogeneity** in the population.

Sleep is another complex behavior. Millions of people suffer from sleep-related disorders. One of these, narcolepsy, is covered later in this chapter. Since many environmental factors can obviously influence sleep (day-night length, stress, work schedule) it has been difficult for researchers to identify a candidate mechanism in humans. However, recall from the earlier discussion that the mating behavior of *Drosophila* is partially determined by the *period* gene. This gene in *Drosophila* has a homologue in humans on chromosome 2. This gene is also called *period*. In a rare type of sleep disorder, called familial advanced sleep phase syndrome (FASPS), people have an altered, and highly predictable sleep-wake cycle. These individuals awaken abruptly at 4:30 AM and fall immediately to sleep at 7:30 PM. The behavior is so predictable that you can set a clock to it. This trait displays an autosomal dominant inheritance pattern in some families, and studies of these families have indicated a single-base substitution in their period gene that may be responsible for the altered phenotype. Unfortunately, some of the families do not possess the substitution, but still display FASPS. So either *period* is only a partial contributor to the phenotype, with other genes yet to be identified, or there are multiple diseases with the same phenotype. Geneticists call this heterogeneity. Heterogeneity and the complex basis of many human behaviors represent a unique challenge for geneticists. In order to determine the genetic basis of a behavioral trait they need to utilize a wide variety of resources, including previous work done in model organisms, such as the fruit fly and mouse, as well as databases of studies performed under controlled conditions. Still, it may be difficult to unravel all of the complex factors that contribute to human behavior.

QUANTITATIVE GENETICS

Geneticists recognize that species are much more complex than the simple expression of single genes. They are also learning that the genome actually interacts with the environment to produce variations in phenotypes. These interactions are the characteristics of a complex trait under the control of many genes (polygenic). Polygenic influence of a complex trait increases the overall number of phenotypic classes significantly. These interactions make a geneticist's work interesting as they research how the genes interact and how much each gene contributes to the phenotype. The previous entry provided a brief introduction to the science of behavioral genetics, or the relationship between the genome and the response of the organism to stimuli in its environment. Many behaviors represent phenotypes that do not necessarily fall into distinct classes. For example, many consider human intelligence to be a behavior. This behavior can be quantified by behavioral scientists using a series of tests that determine the intelligence quotient, or IQ, of the individual. If we administer an IQ test to a randomly selected group of people, we obtain a distribution of IQ scores. This distribution frequently represents a bell-shaped curve (Figure 7.4), with the highest point of the curve representing the most common score of the group being studied. Often, this value is close to being the mean value of the sample, as indicated by the dotted line in Figure 7.4. We can say that a continuous variation exists in the phenotype, with the line in the graph representing the distribution of values in the population. When analyzing this distribution scientists may establish threshold values for classification purposes, such as levels about 140 indicating genius status, or levels below 80 indicating problems in mental development. Yet these are not absolute values, and the interpretation of the phenotypic classes can vary depending on the researcher. This form of phenotypic distribution is a key characteristic of a complex genetic trait.

Complex traits are under the control of more than one gene. However, in the study of genetics there are multiple terms that are used to describe a complex trait. Since the phenotype being examined is under the control of multiple genes, complex traits are commonly called *polygenic traits*. If a trait is under polygenic control, and is also influenced by environmental conditions, as is the case in behavioral genetics (see previous section on "Behavioral Genetics"), the trait is called a *multifactorial trait*. Scientists are recognizing that interactions between the environment and genes are very common. Therefore, it is routine to classify many complex traits as multifactorial. Complex traits are also sometimes said to be *additive*, in that each of the genes involved contribute an additive effect toward the final observed phenotype, although this is not necessarily always the case for every complex trait. For some complex traits the additive effect only occurs when specific alleles of the contributing genes are present in an individual in the right combination, as we will see in the following examples.

The Mendelian characteristics discussed in Chapter 1 frequently focused either on monogenic (single-gene) mechanisms, or in some cases the epistatic interactions of two genes (see "Is Inheritance as Simple as Mendel Suggested?" in Chapter 1). Epistasis is sometimes considered to be a complex trait due to the interactions of the two genes, as was the case with the discussion of Labrador retrievers in the entry "Is Inheritance as Simple as Mendel Suggested?" in Chapter 1. However, recall from these examples that the offspring of the cross possess distinct phenotypes, not

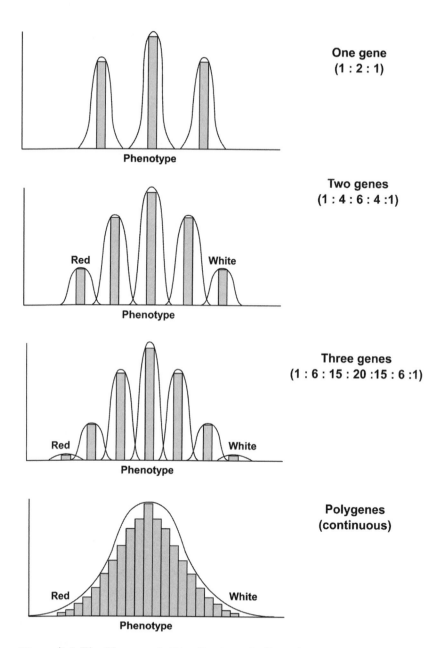

Figure 7.4 The Phenotypic Distribution of a Complex Trait (*Courtesy of Ricochet Productions*).

a continuous variance of the trait. In other cases, epistatic interactions are necessary to produce the phenotypic variation of a quantitative trait. For example, in the metabolic pathways of insecticide detoxification in some insects, including *Drosophila*, epistatic interactions of the enzymes in the pathway are necessary in order to produce high levels of insecticide resistance. The presence of specific detoxification alleles has an additive effect on the resistant phenotype of the individual. The additive nature of genes is a key component of a quantitative trait.

A large part of the research for quantitative geneticists focuses in identifying how many genes are involved in producing the trait being studied. However, before researchers could begin investigating the number of genes involved in producing a quantitative trait, scientists had to first accept that complex traits existed. While Gregor Mendel is clearly recognized as the founding force in the study of the patterns of inheritance in qualitative traits, his laws never address the additive effects of polygenic inheritance. The study of quantitative traits actually predates Mendel and is believed to have begun in the second half of the eighteenth century with work on plant **hybridization** by Joseph Kölreuter.

In one of his experiments, Kölreuter conducted a series of crosses between two strains of tobacco plant (*Nicotiana longiflora*) that differed with regards to height. When dwarf plants were crossed with tall plants, the F_1 generation contained all intermediate height plants. If the experiments had stopped at this point, the intermediate value of the F_1 could be explained as an example of incomplete dominance (see "Incomplete Dominance" in Chapter 1). However, when Kölreuter crossed two of the intermediate plants, the resulting F_2 generation produced plants that varied in height from dwarf to tall and could not easily be placed into distinct phenotypic classes. In effect, Kölreuter had provided one of the first descriptions of a quantitative trait, although he did not elaborate as to the mechanism for the distribution.

In the early twentieth century, following the rediscovery of Mendel's work and a renewed interest into the mechanisms of heredity by the scientific community, there was considerable debate in the scientific world as to whether the quantitative traits in nature could be explained by Mendelian principles. Recall that at this time the nature of the genetic material had yet to be discovered, and a gene simply referred to the hereditary material associated with a trait. Many of these early geneticists believed that all traits were determined by single-gene genetics, which made explaining the mechanism of quantitative traits difficult. In fact, to explain quantitative traits these scientists returned to a blending theory, with the blending of the phenotypes of the parents explaining the quantitative variation of the offspring. Other scientists, including the geneticist William Bateson (1861–1926), contended that all traits were governed by genes, and that the observed quantitative variations from Mendelian genetics could best be explained if multiple genes were contributing to the phenotype.

One of the first scientists to experimentally describe a complex trait was the Swedish botanist Herman Nilsson-Ehle in the early 1900s. Nilsson-Ehle used two strains of wheat (*Triticum aestivum*) that differed in their pigmentation. One strain produced a white grain, while the other produced a red grain. When a red grain plant was crossed with a white grain plant, all of the F_1 generation possessed an intermediate phenotype. However, when the individuals from the F_1 generation were self-fertilized, the resulting F_2 plants demonstrated a spectrum of grain color from red to white.

In most cases when a trait varies continuously in a population it is difficult to sort individuals into distinct phenotypic classes without applying some form of subjective criteria. However, in this case, through careful analysis Nilsson-Ehle was able to identify five phenotypic classes for the F_2 plants. These were red (parental), dark pink, pink, light pink, and white (parental). Furthermore, he noticed that these phenotypes occurred in a 1:4:6:4:1 ratio. A close examination of the 1:4:6:4:1 ratio indicates that the phenotypic classes are all multiples of 1/16. Thus, Nilsson-Ehle recognized that 1:4:6:4:1 ratio was actually a modified 9:3:3:1 ratio, and thus he was dealing with a modified dihybrid cross.

The question was what interaction was occurring between the two genes to produce the 1:4:6:4:1 ratio? A close examination of the Punnett square diagram of this cross (Figure 7.5) indicates that the red parental color is only present in the F_2 offspring that is homozygous dominant (AABB) for both traits, while the white parental color is only present in the homozygous recessive (aabb) offspring. Furthermore, each of these phenotypes only occur in 1/16 of the F_2 generation.

So what about the dark pink, pink, and light pink phenotypes? Notice that the F_2 offspring with the dark red phenotype all have at least three dominant alleles present, and that the phenotype occurs with any combination of dominant alleles, as long as three are present (AABb or AaBB). The same is true of the offspring with the pink phenotype. All of these individuals contain two dominant alleles (AAbb, AaBb, or aaBB). The F_2 offspring with the light pink phenotype all contain a single dominant allele (Aabb or aaBb). The relationship between the number of dominant alleles, phenotype, and phenotypic ratio is illustrated in Figure 7.5. Nilsson-Ehle recognized that there was an additive effect between the dominant alleles, with each dominant allele producing pigment that added to the color of the wheat. Recessive alleles did not contribute, nor did they "subtract" from the phenotype.

The additive effect does not only occur with regards to phenotypes such as pigmentation. In 1957, James Crow (1916–), a *Drosophila* geneticist, demonstrated the additive effect with regards to dichloro-diphenyl-trichloroethane (DDT) resistance in *Drosophila melanogaster*. At the time, DDT was a popular agricultural insecticide and Crow was interested in the evolution of resistance in insect species. Since *Drosophila* were not a target of pest management practices using DDT, Crow had to generate a resistant strain for his work by gradually selecting for rare resistance alleles in a population by slowly increasing DDT concentrations over time to eliminate susceptible individuals.

	ABC	ABc	AbC	CaBC	ABc	aBc	abC	abc
A B C	6	5	5	5	4	4	4	3
A B c	5	4	4	4	3	3	3	2
A b C	5	4	4	4	3	3	3	2
a B C	5	4	4	4	3	3	3	2
A b c	4	3	3	3	2	2	2	1
a B c	4	3	3	3	2	2	2	1
a b C	4	3	3	3	2	2	2	1
a b c	3	2	2	2	1	1	1	0

Figure 7.5 Nilsson-Ehle's Description of a Quantitative Trait in Wheat (*Courtesy of Ricochet Productions*).

Once he had established his resistant strain, Crow crossed resistant and susceptible flies. Males of the F_1 offspring were then crossed back to females of the resistant or susceptible parental strains. This type of cross is called a *backcross*. The F_2 generation produced by the backcross will have differing combinations of susceptible and resistant chromosomes. Using the wealth of morphological mutations known in *Drosophila*, Crow was able to determine which chromosomes were present in each fly. He then exposed groups of flies with the same chromosomal combinations to varying concentrations of DDT to establish the resistance of each chromosomal combination. He determined that the chromosomes containing DDT resistance genes had an additive effect on the phenotype. Later experiments by other researchers have indicated that this type of polygenic response to selection is common in *Drosophila* and other insect species.

What interests us most about the work of these individuals is that their work illustrates an important concept of polygenic analysis. That is, the ability to identify distinct phenotypic classes in a complex trait becomes increasingly difficult as the number of genes contributing to the phenotypic trait increases. In other words, as the number of genes involved increases, the distribution of phenotypes develops the characteristic bell-shaped curve.

As the number of genes increases, the lines between the phenotypic classes become harder to distinguish, until at some point the trait establishes a continuous distribution in the population. Environmental conditions that influence the phenotype, such as with a multifactorial trait, may also blur the boundaries between the phenotypic classes so that fewer genes can still produce a continuous distribution. These types of studies form the basis of the science of quantitative genetics.

GENETICS OF COMPLEX DISEASES

Schizophrenia

Schizophrenia is a complex disease that is characterized by combinations of delusions, hallucinations, problems interacting in society, inability to care for oneself, and frequent depression and anxiety. However, one of the greatest indications of schizophrenia is a decrease in cognitive skills and a fragmentation of thought patterns. The word comes from the Greek words *schizo* (split) and *phrene* (mind), which reflects the fact that the various processes in the brain are no longer functioning together to produce rational thought. It is mistakenly thought of as a multiple personality disorder since the patients may frequently hear "voices," which are actually believed to be either their own thoughts or hallucinations. It is mostly a disease of young adults, with indications of the disease first being displayed between the ages of 18 and 25 in men and 20 to 35 in women. It affects an estimated 1.1 percent of the world's population, or about 51 million individuals. Approximately 2.2 million people are believed to suffer from the disease in the United States. That makes schizophrenia the most common mental illness in the world, ahead of both Alzheimer's and Parkinson's diseases. Twin studies of schizophrenia have provided a strong indication that a genetic component exists for schizophrenia.

Research into the genetics of schizophrenia has focused on the biochemical pathways for two neurotransmitters, dopamine and glutamate. Studies of individuals with schizophrenia have indicated that the dopamine and glutamate levels in the patients are abnormal. Dopamine is a neurotransmitter in the brain that has a wide variety of functions depending on the area of brain in which it is active. It is known to be involved in emotional and motor responses by the body. Glutamate is more of a generalist excitatory neurotransmitter that is known to have a function in the operation of a wide variety of neural pathways. Abnormal neurotransmitter levels are frequently associated with a problem in membrane transport, and thus a significant amount of research has focused on the genes involved in dopamine and glutamate transport.

Fortunately for researchers, there exists information on the heredity of schizophrenia from several large European families, with additional studies using data from North American families, notably the French Canadians. Some of these families have records of schizophrenia going back centuries, supported by well-established pedigrees. Studies of some of these families were instrumental in identifying chromosome 6 as a prime location for schizophrenia genes. Two potential contributors, *NRG1* and *DTNBP1* are located on this chromosome.

Several chromosomal abnormalities have been associated with schizophrenia. These are typically deletions and translocations that disrupt gene function (see "Changes in Chromosome Structure" in Chapter 5). As is frequently the case with chromosomal abnormalities, studies of the deleted and translocated regions of the chromosome can yield insights into possible candidate genes for the complex trait. Such was the case in the identification of *COMT* and *PRODH* as candidate loci for schizophrenia. Both of these genes are located on chromosome 22. Identification of individuals with symptoms of schizophrenia who also possessed a deletion on chromosome 22 led researchers to study these genes as possible candidates. Their status as candidates has been confirmed by SNP-linkage studies (see "Molecular Markers" in Chapter 4). However, it is not believed that deletions or translocations account for a significant percentage of schizophrenia cases.

Linkage studies of the disease to SNPs near candidate loci has clearly demonstrated that schizophrenia is a polygenic trait. The candidate loci were chosen by what is sometimes called a "pathway-driven" strategy. The metabolic pathways for dopamine and glutamate are well known, therefore what several researchers did was to identify SNPs in or near these candidate loci, and then look for linkage of the SNP to schizophrenia. The approach has been successful in identifying several potential contributing genes. Currently there are seven candidate loci that have been linked to schizophrenia, with several others showing promise but lacking concrete linkage data. These genes are scattered throughout the human genome, further reinforcing the fact that schizophrenia is a complex genetic trait.

Narcolepsy

Narcolepsy is a disorder of the sleep-wake cycle that is characterized by excessive sleepiness, hypnagognic hallucinations (dream-like experiences that occur when falling asleep), sleep paralysis, and cataplexy (the sudden loss of control of the voluntary muscles). People who suffer from narcolepsy have an overwhelming desire for sleep, even following sufficient sleep the previous evening. This desire may disrupt normal daily activities such as eating and driving. People with narcolepsy typically also experience disrupted nighttime sleeping patterns. Narcolepsy affects an estimated 200,000 Americans, although fewer than 50,000 of these are diagnosed as having narcolepsy. The disease can occur at any age, and is equally prevalent among the sexes.

From an organismal perspective, narcolepsy is disruption of the normal pattern of events that occur as an individual enters into sleep. Normally, as a person begins to go to sleep, the active brain wave pattern enters into what is called a nonrapid eye movement (nREM) pattern. This is followed by the dream-generating rapid eye movement (REM) pattern, sometimes also called deep sleep, which is characterized as having a more regular pattern of brain waves. In narcolepsy, it appears that the sequence of these events are reversed, with the person entering directly into deep sleep without experiencing nREM brain patterns.

At the cellular level, narcolepsy appears to be associated with the production of specific proteins by the hypothalamus. These proteins are called the orexins (or hypocretins), and three forms (orexin, orexin-a, and orexin-b) are known to be involved in the sleep cycle. Studies of people who have narcolepsy indicate that these proteins are practically missing in the cells of the hypothalamus. In addition to the possible genetic causes of narcolepsy that will be discussed below, there is arising the possibility that the disease is the result of an autoimmune response brought on by an unknown environmental factor. Studies have effectively ruled out structural problems in the brainstem as being a cause of narcolepsy and most research now treats narcolepsy as a neurotransmitter-related disorder.

Studies of monozygotic twins indicate that while a genetic component does exist for the gene, it is actually a complex trait with a strong environmental component. Genetic studies have focused on several candidate loci, including human leukocyte antigen (HLA) and the orexins (hypocretins). A series of studies has indicated linkage to the HLA genes, which are located on chromosome 6. Since HLA has been previously associated with autoimmune diseases, it is thought that some forms of narcolepsy may be caused by an autoimmune response that destroys neurotransmitter-producing cells in the hypothalmus, although this has yet to be conclusively proven by the medical community. However, patients with narcolepsy do not exhibit that characteristic inflammatory response of an autoimmune response, which led researchers to believe that HLA was tightly linked to a narcolepsy gene. Additional studies have shown that this is not the case, and that HLA genes are somehow involved in the disease. The exact role of HLA in narcolepsy still is under investigation.

Studies of human narcolepsy have been aided by successful identification of narcolepsy candidate loci in canines and the mouse. A study of Dobermans in the early 1990s indicated that cataplexy could be induced in canines that were heterozygous for a gene called *carnac-1*. Induction of these narcoleptic symptoms was done using chemical that inhibited the function of specific neurotransmitters in the brain, thus further supporting the concept that narcolepsy is a neurotransmitter-related disease. However, while the *carnac-1* gene in the canine study has been found to be a fully penetrant autosomal recessive gene, the yet to be identified human counterpart is believed to be inherited as an autosomal dominant. The *carnac-1* gene has been mapped to a region of the canine chromosome 12, which has homologous regions on human chromosome 6. This region contains the hypocretin type-2 receptor. Knockout experiments of orexin(hypocretin) in mice (see "Knock-out Mouse" in Chapter 6) indicate that the disruption of this gene produces narcoleptic-type symptoms that are similar to those found in canines and humans. However, it is known that not all cases of narcolepsy can be explained by this mechanism, and the environmental factor that induces the gene has yet to be determined.

Diabetes

Diabetes mellitus is a disease of the pancreas. The pancreas is an interesting organ in the human body in that it acts as a component of the digestive system and the endocrine (hormonal) system. Diabetes is a disease of the hormonal pancreas. Within specialized regions of the pancreas, called the Islets of Langerhans, cells called beta-cells secrete a hormone called insulin. Insulin is a protein-based hormone whose function is to regulate blood glucose levels in the body. In its simplest description, diabetes is the loss of function of insulin in the body, either due to destruction of the beta-cells (type I diabetes), or a developed insensitivity to insulin by the cells of the body (type II diabetes). The disease affects an estimated 18.2 million people in the United States, making it one of the most prevalent diseases in our population. 1.3 million people are diagnosed with diabetes each year, with over 200,000 people dying each year either directly from diabetes, or from complications associated with the disease. Diabetes is linked to an increased rate of heart and kidney disease, stroke, blindness, and nervous system damage. In general, it can affect people of any age, sex or race, although some races do have a higher prevalence for the disease. Diabetes is a complex trait in that it is known to be under polygenic control with a substantial environmental influence. The two forms of diabetes differ in their genetic mechanisms and the influence of the environment.

Insulin-dependent diabetes mellitus (IDDM), also called Type 1 diabetes and juvenile diabetes, is characterized by the destruction of the beta cells in the pancreas. This usually occurs as the result of an autoimmune response, although both infections (viral and bacterial) and diet have also been implicated.

Studies of the human genome have identified 18 regions that are associated with IDDM. These regions are called IDDM1 through IDDM18. The genes contributing to IDDM have not been conclusively identified in some of these regions, and may most likely be associated with an increased risk of disease resulting in destruction of the beta cells. Genes in other locations are most likely minor genes with an additive role in increasing the risk of diabetes in response to a specific environmental factor. However, two of the regions (IDDM1 and IDDM2) contain genes of special interest.

IDDM1 occupies a region on the short arm of chromosome 6. Contained within this region are the genes that the body utilizes to manufacture the cellular identification tags (glycoproteins) that the cell uses to identify itself to the immune system. These glycoproteins are part of the major histocompatability complex (MHC), a series of complex markers located on the cell membrane. The proteins are encoded by HLA genes. Mutations in the HLA genes may result in an autoimmune response, one of the suspected causes of Type 1 diabetes. However, it is important to note that the IDDM locus also contains other genes that increase susceptibility to diabetes, although their identity is still being determined.

IDDM2 contains the most likely candidate for diabetes, the insulin gene. The gene for insulin is located on chromosome 11. As mentioned previously, insulin is a hormone produced by the pancreas. The insulin protein is actually made from two peptides called A and B chains. These are linked together by sulfide bonds to produce a functioning insulin molecule. Several polymorphisms exist near the insulin gene that allow specific variants of the gene to be identified in populations. These polymorphisms include SNPs and VNTRs (see "Molecular Markers" in Chapter 4). It is possible that IDDM2 contains other yet to be identified genes associated with diabetes. Genes in the IDDM locus account for approximately 10 percent of the cases of Type 1 diabetes.

The second form of diabetes is non-insulin-dependent diabetes mellitus (NIDDM). It is also called Type 2 diabetes or adult-onset diabetes and is the most common form of the disease. This form of disease is characterized by a lack of sensitivity to insulin by the cells of the body, and a resulting increase in blood glucose levels. This decreased sensitivity is due primarily to the fact that adipose tissue does not respond well to insulin, so as a person's weight increases, they acquire a reduced response to insulin being produced by the pancreas. Type 2 diabetes can progress to Type 1 diabetes. The disease is not limited to adults, and is increasingly found in overweight children in the Western world.

The environment of exercise and diet play a major role in the development of Type 2 diabetes. However, there are a number of genes that are known to increase susceptibility to the disease. These are scattered throughout the genome and are believed to have an additive effect in producing the diabetic phenotype. There is little doubt in the research community that this is not the full extent of the candidate genes producing increased susceptibility to Type 2 diabetes. It is important to note that although these genes increase susceptibility, in most cases this form of diabetes can be controlled by regulating the environmental factors of diet and exercise.

Retinitis Pigmentosa

The vision systems of mammals are complex, and problems associated with these systems can occur in the eyes or the interpretation centers of the brain. With regard to humans, there are many forms of blindness. Some of these may be caused by tumors and cancer (such as retinoblastoma—see "Cancer Genetics" in this chapter) and destroy the sight of the person permanently. Other vision disorders limit the sight of the person, either with regard to the field of view or the wavelengths of light that are detected. One form of blindness is called retinitis pigmentosa (RP). Unlike color-blindness, RP is considered to be a complex trait since it may be caused by defects in several genes, and appears to have an environmental component.

RP is the most common hereditary cause of human blindness and is known to affect an estimated 200,000 Americans. The disease begins with night blindness and loss of peripheral vision. It can progress to a drastic restriction of the visual field , or a complete loss of vision. There are no known cures for RP. Inheritance patterns for the disease indicate that some forms are X-linked, while other forms exhibit autosomal dominant and autosomal recessive inheritance. In addition, there is one mitochondrial mutation that is known to cause RP. There are 39 known genes that can contribute to RP, and yet these only account for a small portion of the known cases, suggesting that there are many other, yet to be identified, genetic mechanisms. The disease has a high degree of *heterogeneity*, which is believed to have complicated the identification of all of the genetic mechanisms. The genes that cause RP have been identified in the pathways that process visual signals, the recycling of visual pigments, and in the structure of the photoreceptors. Some researchers believe that unidentified environmental factors may also be producing the disease.

There are five X-linked genes that are known to be associated with RP. Researchers have identified two of these genes (*RP2* and *RPGR*) that map to different regions of the X chromosome. Three other unidentified loci have been mapped to a third region of the X chromosome. The *RP2* gene is most likely involved with the folding of tubulin, a structural protein, and *RPGR* may be involved with photoreceptor cilia function, although the exact function of both of these gene products is under investigation. Within these genes researchers have identified 43 unique mutations that cause the disease.

The autosomal recessive form of RP is the rarest form of the disease. Twenty-one candidate loci contributing to this form of RP have been identified on chromosomes 1, 2, 3, 4, 5, 6, 10, 15, and 16. One of these mutations, in the *CERKL* gene, has been mapped to human chromosome 2 using SNPs and microsatellites (see "Molecular Markers" in Chapter 4). This gene is believed to be involved with the process of apoptosis, or cell death. The mutation in this gene produces an early stop codon. The researchers who identified this gene speculate that the mutation may make the photoreceptors most susceptible to apoptosis and result in premature cell death. However, this mutation only accounts for a small percent (<1%) of all cases of RP.

Autosomal dominant RP has been mapped to chromosomes 1, 3, 6, 7, 8, 11, 14, 17, and 19. However, four genes are responsible for the majority of the dominant forms of this disease. The *RHO* gene (chromosome 3) is a rhodopsin gene. Over 100 mutations that can cause RP have been detected in this gene. The *RP1* gene (chromosome 8) product is an oxygen-regulated protein involved in photoreceptor function. Mutations in *RDS* (chromosome 6), a gene that encodes a protein called peripherin, is the third leading cause of autosomal dominant RP. The gene *PRPF31* (chromosome 19) is involved in pre-mRNA gene splicing, and accounts for about 15 percent of all cases of autosomal dominant RP. As noted, these genes only account for approximately 50 percent of all cases of RP. It has been shown that some individuals may contain multiple mutations, thus further complicating the identification of the genetic mechanisms of this complex disease.

GLOSSARY

Alleles. A variation of a gene. An allele is due to a change in the sequence of nucleotides at the molecular level. This change may often be as minor as a single nucleotide substitution.

Aneuploidy. A condition where the total number of chromosomes in a cell (or organism) is not an exact multiple of the chromosome number (n). Trisomy ($2n + 1$) and monosomy ($2n - 1$) are examples.

Antibodies. A protein produced by the immune system that recognizes specific molecules and marks them for destruction.

Anticodon. A three-nucleotide region of the tRNA that recognizes a specific codon on the mRNA for protein synthesis.

Apoptosis. The process of programmed cell death. Apoptosis is an important aspect of normal tissue development and aging.

Association study. In the study of human disease, this is the comparison between an individual who has a specific disease and one who is free of the disease. Association studies are often conducted using genetic markers such as SNPs, RFLPs, or VNTRs.

Bacteriophage. A form of a virus that infects bacteria. There are often used in genetic studies due to the small size of their genome.

Blastocyst. An early embryonic stage that consists of a hollow ball of cells. These cells are undifferentiated and are a form of stem cell.

Carcinogen. A chemical or form of radiation that induces mutations associated with the formation of cancer.

Cell cycle. A series of stages that somatic cells pass through prior to cell division.

Central dogma. A term used to describe the process of gene expression. It represents the flow of information from the DNA to mRNA and then protein by the processes of transcription and translation.

Centromere. A region of highly condensed DNA on a chromosome that plays an important role in the separation of the genetic material during cell division. The location of the centromere is used to partially identify chromosomes.

Chimera. Any organism that has been produced by combining the cells from two different organisms.

Chromatin. A DNA-protein complex that is responsible for compacting the genetic material in a cell.

Chromosomes. A structure composed of DNA and proteins that contains the genes of an organism. The number of chromosomes and their length varies between species. In the bacteria, chromosomes are often circular.

Codominance. When two alleles act in a dominant manner in a cell. In this case, the gene product of both genes is expressed in the organism.

Codon. A group of three nucleotides in the mRNA that code for a specific amino acid.

Consensus sequence. In comparisons of DNA sequences, this is the most common nucleotide found at each position in the sequence.

Cytoplasm. The semifluid interior of a cell. In a eukaryotic cell, this does not include the spaces within the membranes of the organelles.

Dideoxy method. A form of DNA sequencing that uses chemically modified nucleotides, called dideoxynucleotides (ddNTPs), to interrupt DNA replication. These nucleotides are frequently labeled with radioisotopes to provide a visual display of the results.

Diploid. A cell, tissue, or organism that contains two copies of each type of chromosome. Diploid is often represented as *2n*, where *n* equals the chromosome number of the species.

DNA library. In the process of DNA cloning this represents a group of bacteria that hold the fragmented pieces of a genome. cDNA libraries only contain the expressed genes.

Dominance. A term, first used by Gregor Mendel, to describe a trait that masks the expression of a second trait.

Endonucleases. Enzymes that degrade nucleic acids within a cell.

Enzyme. A catalyst of a metabolic reaction. Enzymes accelerate chemical reactions. Most enzymes are proteins, however, some RNAs are known to have enzymatic properties.

Epistatis. A term used to describe two or more genes that interact in order to produce a phenotype.

Euploid. A cell having the correct number of chromosomes. For a diploid organism, a euploid cell would be *2n*, where *n* is the chromosome number of the species.

Exons. The regions of a gene that contain the information for the formation of a popypeptide.

Fluorescent in situ hybridization. Also known as FISH, this is a procedure that uses fluorescent tags to either identify chromosomes or mark the physical location of a gene on a chromosome.

Free radicals. A highly reactive atom or compound that can damage DNA, proteins, or other organic compounds in a cell. Ozone is an example of a free radical.

Functional group. The portion of an organic molecule that is largely responsible for its chemical characteristics. Examples are phosphate group, hydroxyl groups, and amine groups.

Gene expression. A multistage process that involves transcription, RNA editing (in eukaryotes), and translation to produce a protein.

Gene families. A group of genes that have similar nucleotide sequences and function in the genome of a species. These genes are usually found in clusters along a chromosome and are believed to be produced by duplication and unequal recombination.

Gene pool. The sum of the genes in a population.

Genes. The genetic instructions for the production of a polypeptide or functional RNA (such as tRNA and mRNA). Genes consist of regulatory regions and coding regions.

Genetic drift. The influence of chance events on the frequency of alleles in the gene pool of a population.

Genetically modified organism. An organism whose DNA has been altered using recombinant techniques in order to improve a specific trait.

Genome. This term can either be used to describe the sum of all of the genetic information in the chromosomes of a cell or the sum of all of the genes in a cell.

Genotype. The combination of alleles in a cell or organism. In a diploid cell the genotype consists of two alleles.

Germ cells. Those cells which are responsible for producing the gametes. It is the genetic information in these cells that are passed on from generation to generation.

Glycoprotein. A protein that has a sugar (carbohydrate) group attached. They are commonly found in the cell membrane where they are involved in cell identification.

Haploid. A cell or organism that contains half the normal complement of chromosomes. Haploid cells are produced by meiosis and are used for sexual reproduction.

Hardy-Weinberg equation. An algebraic formula that defines the frequency of alleles in a stable, or unchanging, population.

Hemoglobin. The oxygen-carrying protein of blood. Hemoglobin contains four subunits and iron atoms.

Heritability. A ratio that determines how much of a complex trait is due to genetics and how much due to the environment. A heritability value of 1.0 indicates complete genetic control, while a value of 0.0 indicates complete environmental control.

Heterogeneity. Similar phenotypes that are caused by different genetic factors.

Heterozygous. The term used to describe an individual who has two different alleles for a trait.

Homologous. Chromosomes that are practically identical in the alleles that they contain. Homologous chromosomes may still contain minor variations in the nucleotide sequences.

Homologous recombination. Recombination that involves two similar sequences of DNA. In gene therapy, this is the replacement of one gene with another gene that has a similar sequence.

Homozygous. The term used to describe an individual who has two identical alleles for a trait.

Human Genome Project. A project begun in the 1990s to sequence all of the chromosomes in the human genome.

Hybrid vigor. Also known as heterosis, this represents the condition by which a heterozygote is better adapted to a specific environment than either the homozygous dominant or homozygous recessive combination.

Hybridization. The pairing of complementary nucleotide sequences. The use of RNA or DNA probes in molecular biology makes use of RNA–DNA or DNA–DNA hybridization.

Hydrogen bonds. A weak chemical attraction between the hydrogen atoms of one molecule and a negative portion of another molecule. While hydrogen bonds are individually very weak, they are usually present in high enough numbers to present a very strong attractive force.

Hydrophilic. Any molecule or compound that attracts, or is attracted to, water.

Hydrophobic. Any molecule or compounds that repulses, or is repulsed by, water.

Incomplete dominance. When the phenotype of a dominant allele does not completely mask the phenotype of a recessive allele.

Introns. Also called intervening sequences, introns are regions of the gene that do not contain coding information for a polypeptide. Introns are removed by RNA splicing.

Isotope. Atoms that have the same atomic number but different atomic masses. Isotopes are caused by changes in the number of neutrons in the nucleus. Some isotopes are unstable and are termed radioactive.

Karyotype. A photograph or computer image of the chromosomes in the genome of a cell. Also called ideograms, karyotypes are used to study changes in chromosomal structure or number.

Knock-out mouse. A transgenic mouse in which one of the genes for a trait has been disabled in order to generate a strain of mice that are homozygous for a specific defective gene.

Lateral gene transfer. The movement of genes from one species to another, usually by the use of an intermediate such as a virus.

Law of independent assortment. Also known as Mendel's second law of heredity, this law states that factors for different traits assort independently from one another.

Law of segregation. Also known as Mendel's first law of heredity, this law states that alleles for a given trait separate from one another during the formation of gametes.

Lipoprotein. A protein with a lipid (fat) group attached. These are often used as transport molecules to move fats and cholesterol through the circulatory system.

Liposome. A collection of lipids that forms a small hollow ball. Liposomes are often used to introduce lipid-soluble compounds into a cell.

Locus. The physical location of a gene on a chromosome. This may be determined by information provided by studies of recombination, or by physically mapping the gene to the chromosome.

Macroevolution. The study of long-term trends and patterns in evolution.

Meiosis. Also known as reduction division, this is a form of cell division that produces daughter cells with one half of the chromosome number of the parent cell. In animals, meiosis is the first step in the production of gametes.

Microarrays. Also called "gene chips," these contain thousands of genes on a small silicon slide. They are frequently used to determine patterns of gene expression under specific environmental conditions.

Microevolution. The change in allele frequencies of a population over time. Studies of microevolution usually focus on changes over relatively short periods of time, such as a few generations.

Microsatellites. A series of one, two, or three nucleotides that are repeated in tandem in the genome. The sequence may be repeated up to 100 times. Microsatellites are a form of variable number of tandem repeats (VNTRs).

Minisatellites. A series of 20–100 nucleotides that are repeated in tandem up to several thousand times. Minisatellites are a form of variable number of tandem repeats (VNTRs)

Mitosis. A form of asexual reproduction that produces cells that are genetically identical and contain the same number of chromosomes as the parent cell.

Monosomy. A cell or organism that is missing one copy of a chromosome.

Monozygotic twins. Also called identical twins, monozygotic twins are formed from the same zygote, and therefore have almost exactly the same genetic information.

Multifactorial. A trait, usually polygenic, that is under the influence of environmental factors. This is sometimes called a complex trait.

Mutagens. A chemical or other agent that catalyzes a mutation in the nucleotide sequence of DNA.

Mutation. A change in the nucleotide sequence of the DNA.

Mutation frequency. The incidence of a specific mutation within a given population.

Mutation rate. The probability that a mutation will occur during a specific unit of time. In population genetics the reference time is usually the generation time of the organism.

Nondisjunction. A failure of the chromosomes to separate correctly during anaphase of either mitosis or meiosis. This can result in either an aneuploid or polyploidy condition for the cell.

Nonrecombinants. The term used to describe a chromosome that is a replicate of the original parental chromosome. Nonrecombinants have not undergone crossing over in prophase I of meiosis.

Nuclear pores. Protein channels in the nuclear membrane that allow the movement of material in and out of the nucleus.

Nucleolus. A region of the nucleus in which the ribosomal subunits are assembled from rRNA and ribosomal proteins. In cell biology, the nucleolus will appear as a darkened area within the nucleus.

Nucleotides. The basic building blocks of DNA and RNA. Each nucleotide contains a five-carbon sugar, a phosphate group, and a nitrogenous base. Nucleotides are also used as metabolic molecules, such as the energy carrier adenosine triphosphate (ATP).

Oligonucleotide. A short sequence of nucleotides. They are often used as primers in the polymerase chain reaction or as probes in Southern blots.

Operator. A regulatory region of an operon. The operator is the site where regulatory proteins bind to influence the expression of the genes in the operon.

Operon. An arrangement of genes in bacteria. An operon contains a group of genes that are transcribed together, but are under the control of a single regulatory region.

Palindrome. A series of nucleotides that is the same when read in both directions.

Particulate theory of inheritance. An early theory of inheritance that states that a distinct unit carries the information for a trait from generation to generation. That unit is now known as the gene.

Pedigree. A graphic representation of family inheritance. Pedigrees are often used to determine the basis of inheritance (dominant or recessive) and whether a given trait is sex-linked.

Photolithography. A photographic technique in which a pattern is transferred to a substrate using chemical etching. Photolithography is used by the electronics industry for the production of integrated circuits.

Plasmid. A small, circular piece of DNA found in many prokaryotes and some eukaryotes. Plasmids often contain beneficial genes, such as those conferring resistance to antibiotics.

Pleiotrophic. A gene that produces many different traits in an organism. Often these traits appear to be unrelated.

Polygenic. A trait that is under the control of more than one gene.

Polymerase chain reaction. Commonly called PCR, this is a chemical reaction that mimics that process of DNA replication in a test tube. It is used by geneticists to amplify specific sequences of DNA for study.

Polymorphic. A gene that has more than two alleles in a given population. In order for an allele to be considered polymorphic it must be present in at least 1 percent of the population.

Polypeptide. A sequence of amino acids encoded by a gene. Polypeptides are the precursors to proteins, the working molecules of the cell.

Polyploidy. A condition by which a cell or organism has an extra set, or sets, of chromosomes. Examples are triploidy (3 sets), tetrapolidy (4 sets), and hexaploidy (6 sets).

Polysaccharide. A complex carbohydrate consisting of long chains of sugar molecules.

Population. A group of individuals of the same species occupying a given geographic area.

Probability. The likelihood that an event will occur. Probability values range from 1.0 to 0.0, with 1.0 indicating a 100 percent chance that the event will occur.

Probe. A sequence of nucleotides (or amino acids) that have been labeled with either a radioactive or fluorescent tag. The probe will hybridize with specific sequences of nucleotides, thus indicating their position on a gel, blot, or cloning plate.

Prokaryotic. An organism that lacks a nucleus and membrane-bound organelles. Commonly called a bacteria.

Promoter. The region of a gene at which the RNA polymerase binds to begin transcription. The promoter is a regulatory region, and does not contain coding information for the polypeptide.

Proofreading. The ability of certain molecules, namely the DNA polymerases, to recognize and correct errors in the genetic information.

Proteome. The sum of all of the proteins produced by a cell. In eukaryotic species, the size of the proteome is usually much greater than the number of genes in the genome.

Pseudogene. A "false" gene. Pseudogenes typically represent genes that have been inactivated over evolutionary time frames by mutations or transposons.

Recombinants. The term used to describe a chromosome that is a combination of the original parental chromosomes. Production of a recombinant occurs during the initial phases of meiosis.

Recombinant DNA. A DNA molecule that has been made in the lab by combining two different sequences of DNA.

Restriction enzymes. Enzymes that are capable of breaking the phosphodiester bonds between nucleotides. Restriction enzymes recognize specific sequences of nucleotides and are present in some organisms to protect against foreign DNA. They are also called *restriction endonucleases*.

Reverse transcriptase. An enzyme found in some viruses that allows the formation of a DNA strand from a RNA template.

RFLP. A restriction fragment length polymorphism. A form of molecular marker produced by variations in the length of DNA fragments formed when a stretch of DNA is cut using a restriction enzyme.

Ribozyme. The name given to an RNA molecule that has catalytic characteristics. The ribosome is an example of a ribozyme.

RNA editing. Following transcription, the process by which the ends of the pre-mRNA are modified and introns sequences removed.

RNA polymerase. The protein complex that is responsible for copying the coding regions of a gene during transcription.

RNA splicing. The removal of introns from a transcribed nucleotide sequence to produce a functional mRNA.

Scientific method. A system used by scientists that involves the collection of data, the development of hypotheses to explain the data, and the use of experimentation to test the validity of the hypotheses.

Sex-linked traits. A trait or condition that is inherited with one of the sex chromosomes. The majority of sex-linked traits in humans are due to genes on the X chromosome.

Somatic cells. The cells of the body that are not involved in the production of gametes.

Spliceosome. A protein complex that is involved in the removal of introns in eukaryotic organisms.

Structural gene. A gene that when transcribed produces a protein for the cell.

Telomerase. The enzyme that is responsible for replicating the telomeres.

Telomeres. The ends of a eukaryotic chromosome. Telomeres typically do not contain genes, and are comprised of highly repetitive sequences of DNA.

Teratogen. A chemical or form of radiation that causes a birth defect. Teratogens are often mutagens because they cause a mutation in a developmental gene.

Transcription. The copying of the DNA to produce an mRNA molecule. Transcription is the first stage of gene expression and involves the action of the RNA polymerase molecule.

Transgenic. An organism that contains recombinant DNA from two different species. The term usually means that all of the cells of the organism contain the recombinant DNA.

Translation. The process by which the mRNA is read to form a polypeptide chain. Translation occurs at the ribosome.

Transposition. The movement of a transposon within the genome. Transposition is facilitated by the enzymes transposase and resolvase.

Trisomy. A cell or organism that has an extra copy of one chromosome.

Two-hit hypothesis. This states that in order inactivate the tumor-suppressing genes in a cell, both copies of the gene must accumulate loss-of-function mutations.

Variable number of tandem repeats. A series of tandem nucleotide repeats in the genome. The number of repeats is highly polymorphic in a population and may be used as a genetic marker or as the basis for DNA fingerprinting.

Vector. In the study of genetics this term is used to identify an agent that can deliver DNA sequences inside of a cell. Viruses and plasmids are common vectors in molecular biology.

Virus. A nonliving, infectious particle composed of proteins and nucleic acids. Viruses hijack the machinery of a cell in order to manufacture more viruses.

Vitamin. An organic compound that serves as an assistant to a metabolic pathway.

Wild-type. The most common allele of a trait in a natural population. In most cases, the wild-type allele is dominant to recessive mutations. The term is used widely in the study of genetics in *Drosophila*.

References and Resources

PRINT

Asimov, Isaac. *Asimov's Chronology of Science & Discovery.* New York: HarperCollins Publishers, 1994.

Brody, Jonathan R., and Scott E. Kern. "History and Principles of Conductive Media for Standard DNA Electrophoresis." *Analytical Biochemistry* 333 (2004): 1–13.

Brooker, Robert. *Genetics: Analysis and Principles,* 2nd ed. Dubuque, IA: McGraw-Hill Higher Education, 2005.

Bud, Robert. *The Uses of Life: A History of Biotechnology.* Cambridge, UK: Cambridge University Press, 1993.

Bynum, W. F., E. J. Browne, and R. Porter. *The Dictionary of the History of Science.* Princeton, NJ: Princeton University Press, 1981.

Carlson, E. A. *Mendel's Legacy: The Origin of Classical Genetics.* Cold Spring Harbor, NY: Cold Spring Harbor Laboratory Press, 2004.

Cech, Robert. "RNA as an Enzyme." *Scientific American* 255(5) (November 1986): 64–73.

Fuller, Watson. "Who said 'Helix'?" *Nature* 424(6951) (2003): 876–878.

Gribben, John. *The Scientists: A History of Science Told through the Lives of Its Greatest Inventors.* New York: Random House, 2002.

Guttman, B., A. Griffiths, D. Suzuki, and T. Cullis. *Genetics: A Beginner's Guide.* Oxford, UK: Oneworld Publications, 2002.

Hartwell, L. H., L. Hood, M. L. Goldberg, A. E. Reynolds, L. M. Silver, and R. C. Veres. *Genetics: From Genes to Genomes,* 3rd ed. Boston, MA: McGraw-Hill Higher Education, 2008.

Holmes, Frederic L. *Meselson, Stahl and the Replication of DNA.* New Haven, CT: Yale University Press, 2001.

Jaenicke-Després, V., E. Buckler, B. Smith, M. T. Gilbert, A. Cooper, J. Doebley, and Svante Pääbo. "Early Allelic Selection in Maize as Revealed by Ancient DNA." *Science* 302(5648) (November 2003): 1206–1209.

Kay, Lily E. *Who Wrote the Book of Life? A History of the Genetic Code.* Stanford, CA: Stanford University Press, 2000.

Keller, Evelyn Fox. *The Century of the Gene.* Cambridge, MA: Harvard University Press, 2000.

Krebs, Robert E. *Scientific Laws, Principles, and Theories: A Reference Guide*. Westport, CT: Greenwood Publishing, 2001.

Lagerkvist, Ulf. *DNA Pioneers and Their Legacy*. New Haven, CT: Yale University Press, 1998.

Lewis, R. *Human Genetics: Concepts and Applications*, 6th ed. Boston, MA: McGraw-Hill Higher Education, 2005.

Maas, Werner. *Gene Action: A Historical Account*. New York: Oxford University Press, 2001.

Mendel, Gregor. Experiments in Plant Hybridization. In *Classis Papers in Genetics*, James A. Peters (ed.). Englewood Cliffs, NJ: Prentice-Hall, Inc., 1959.

Miller, Orlando J., and Eeva Therman. *Human Chromosomes*, 4th ed. New York: Springer-Verlag, 2001.

Monaghan, Pat, and Mark F. Haussmann. "Do Telomere Dynamics Link Lifestyle and Lifespan?" *Trends in Ecology and Evolution* 21(1) (January 2006): 47–53.

Morange, Michel. *A History of Molecular Biology*. Cambridge, MA: Harvard University Press, 1998.

Mullis, Kary. "The Unusual Origin of the Polymerase Chain Reaction." *Scientific American* (April 1990): 55–65.

Olby, Robert. *The Path to the Double Helix*. Seattle: University of Washington Press, 1974.

Pierce, Benjamin A. *Genetics: A Conceptual Approach*. New York: W. H. Freeman and Company, 2003.

Rabinow, Paul. *Making PCR: A History of Biotechnology*. Chicago, IL: University of Chicago Press, 1996.

Russell, Peter J. i*Genetics*. San Francisco, CA: Benjamin-Cummings, 2002.

Sen, G. L., and H. M. Blau. "A Brief History of RNAi: The Silence of the Genes." *FASEB* 20 (2006): 1293–1299.

Serafini, Anthony. *The Epic History of Biology*. Cambridge, MA: Perseus Publishing, 1993.

Sokolov, Raymond. "The Unknown Bioengineers." *Natural History* 102(10) (October 1993): 104–108.

Wallace, Bruce. *The Search for the Gene*. Ithaca, NY: Cornell University Press, 1992.

Watson, J. D. *The Double Helix: A Personal Account of the Discovery of the Structure of DNA*. New York: Atheneum, 1968.

Watson, J. D., and F. H. C. Crick. "A Structure for Deoxyribose Nucleic Acids." *Nature* 171 (1953): 737–738.

Windelspecht, Michael. *Groundbreaking Scientific Experiments, Inventions & Discoveries of the 19th Century*. Westport, CT: Greenwood Publishing, 2003.

Wright, Robert. "James Watson and Francis Crick." *Time* 153(12) (1999): 172–176.

Zohary, Daniel, and Maria Hopf. *Domestication of Plants in the Old World*. Oxford, UK: Clarendon Press, 1988.

WEB

Historical

- Mendel Museum (www.mendel-museum.org). This site is maintained by the Museum of Genetics at the Abby of St. Thomas in Brno, Czech Republic. It contains a wealth of information on Mendel and the environment he worked in during the mid-1900s. It also contains excellent images of Mendel, as well as information on the science he conducted on pea plants.

- Nobel Prize Foundation (nobelprize.org). The study of the history of DNA and the gene has led to a number of Nobel Prizes in chemistry or medicine. This site provides background information on the recipients, as well as transcripts of their acceptance speeches and descriptions of their work.

- Cambridge Physics (www.cambridgephysics.com). The site contains a history of the major discoveries in the Cavendish Laboratory at Cambridge University, the site of Watson and Crick's discovery of DNA structure. The internal link to Watson and Crick's work takes you through the major events leading up to their discovery.

- DNA Interactive (www.dnai.org/). This site contains links to a timeline of important events in the study of DNA.

- MendelWeb (www.mendelweb.org/Mendel.html). This site contains an English version of Gregor Mendel's manuscript. The manuscript contains useful hyperlinks to a glossary to assist with terminology.

- The Mechanism of Heredity (www.cshl.edu/History/100years-t14.html). This site provides a narrative review of the work done at the beginning of the twentieth century that led to the proof of the chromosome theory of inheritance. Photos of Thomas Hunt Morgan, Calvin Bridges, and Alfred Sturtevant are provided.

- Time 100 (www.time.com/time/time100/scientist/profile/watsoncrick.html). Time magazine named James Watson and Francis Crick one of the 100 most influential people of the twentieth century. This article gives a brief history of the events leading up to the discovery of DNA structure.

- Polymerase Chain Reaction at PCRlinks.com. (www.pcrlinks.com/generalities/history.htm). Information on how the polymerase chain reaction was developed in the mid-1980s.

Genomics

- Human Genome Resources (www.ncbi.nlm.nih.gov/genome/guide/human/). Hosted by the National Center for Biotechnology Information (NCBI) and Johns Hopkins University, this site serves as a portal to a wealth of information on the human genome. Contained within the site is the Online Mendelian Inheritance in Man (OMIM) database where you can examine details about known genetic causes for a variety of diseases.

- Department of Energy's Genome Project (www.doegenomes.org/). Initially designed as the Internet site for research on the human genome, this Web site now includes links to a wide variety of genome-related projects funded by the U.S. government. Links are included to research in genomics and advances in the study of microbial genomes.

- Virtual Library of Genetics (www.ornl.gov/sci/techresources/Human_Genome/genetics.shtml). This site is hosted by the U.S. Department of Energy and contains a vast array of information on genetics. There are links to genomics projects in a variety of organisms, as well as maps and information on the human chromosomes.

- University of Santa Cruz Genome Resources (genome.ucsc.edu/). A site that contains access to a number of genomes, including humans. Follow the link "Genome Browser" to begin your search.
- EBI/Sanger Center (www.ensembl.org). An easy-to-navigate site that provides links to a large number of genomes currently being studied by a number of different genetics labs.
- NCBI Human Genome Resources (www.ncbi.nlm.nih.gov/genome/guide/). This site contains some useful tools for genomic research. From the homepage it is possible to examine all of the information on a single human chromosome. Additional useful links on this page include those for electronic PCR and a gateway into the Online Mendelian Inheritance in Man (OMIM).
- DOE Joint Genome Institute (genome.jgi-psf.org/). Contains links to projects currently in progress in the labs associated with the Department of Energy genome projects, as well as gateways to sequence information on a wide range of genomes.

Diseases and Disorders

- Online Mendelian Inheritance in Man (OMIM; www.ncbi.nlm.nih.gov/entrez/query.fcgi?db=OMIM). This site is hosted by the National Society for Biotechnology Information (NCBI). It contains a vast database of information on human genes and genetic disorders.
- Genetics Disorders Library (gslc.genetics.utah.edu/units/disorders/whataregd/). Hosted by the Genetics Science Learning Center at the University of Utah, this site contains information on a variety of human genetic disorders. Of special interest are the links for educators which provide lesson material for use in the classroom.
- Sickle Cell Information Center at Emory University (www.scinfo.org/). Contains many useful links to information on sickle-cell disease, including information for patients and health-care providers.

Dictionaries and Encyclopedias

- Talking Glossary of Genetic Terms. (www.genome.gov/glossary.cfm). This site is part of the National Human Genome Research Institute sponsored by the National Institutes of Health. Although the site is not a comprehensive list of genetic terms, it does contain short audio files explaining the terms. Explanations are available in Spanish for most terms.
- Glossarist. (www.glossarist.com/glossaries/science/life-sciences/genetics.asp). This site serves as a portal to a variety of online dictionaries in genetics.

Educational

- Genetics Science Learning Center (gslc.genetics.utah.edu/). This site, sponsored by the University of Utah, contains lesson plans for teachers and at-home experiments. It also contains links to hot topics in genetics, such as cloning and pharmacogenomics.
- Genetics Home Reference (ghr.nlm.nih.gov/). This site is maintained by the National Library of Medicine (NLM). It contains a "Handbook of Genetics" that is

designed to help people understand the concepts of genetics. Also included are human genes that are currently in the news, and a comprehensive glossary of genetic terms.

- DNA to RNA to Protein at NobelPrize.org (nobelprize.org/educational_games/ medicine/dna/index.html). This site provides an overview of the movement of information from DNA to a functional protein. By navigating the links it is possible to obtain more information on the processes of DNA replication, transcription, and translation.
- Science Odyssey: DNA Workshop at PBS.org (www.pbs.org/wgbh/aso/tryit/ dna/). This site contains an interactive animation that allows you to manipulate DNA to understand the processes of DNA replication, transcription, and translation.
- The Biology Project at the University of Arizona (www.biology.arizona.edu/ mendelian_genetics/mendelian_genetics.html). This site contains several examples of monohybrid, dihybrid, and sex-linked crosses.
- The Cell Cycle & Mitosis Tutorial, The Biology Project (www.biology.arizona. edu/cell_bio/tutorials/cell_cycle/main.html). This site provides an overview of the cell cycle and mitosis. The animations provide a useful method of understanding the cell cycle.

Professional Associations

- Genetics Education Center (www.kumc.edu/gec/prof/soclist.html). This site is part of a larger Web site on genetics hosted by the University of Kansas Medical Center. This page contains a worldwide listing of the major genetics-related professional organizations.
- American Society of Human Genetics (www.ashg.org/genetics/ashg/ashgmenu. htm). While primarily designed for genetics professionals, the educational links contain useful information for those wishing to understand current research in genetics.
- Genetics Society of America (genetics.faseb.org/genetics/g-gsa/index.shtml). One of the larger professional associations for researchers of genetics. The menu option on teaching genetics contains a number of links that may be useful for nonscientists.

INDEX

About the Author

MICHAEL WINDELSPECHT is Assistant Professor of Biology at Appalachian State University. He is the author or coauthor of numerous works for Greenwood, including *The Digestive System* and *The Lymphatic System* in the "Human Body Systems" series.